Lecture Notes in Applied and Computational Mechanics

Volume 24

Series Editors

Prof. Dr.-Ing. Friedrich Pfeiffer
Prof. Dr.-Ing. Peter Wriggers

Lecture Notes in Applied and Computational Mechanics

Edited by F. Pfeiffer and P. Wriggers

Vol. 23: Frémond M., Maceri F. (Eds.)
Mechanical Modelling and Computational Issues in Civil Engineering
400 p. 2005 [3-540-25567-3]

Vol. 22: Chang C.H.
Mechanics of Elastic Structures with Inclined Members: Analysis of Vibration, Buckling and Bending of X-Braced Frames and Conical Shells
190 p. 2004 [3-540-24384-4]

Vol. 21: Hinkelmann R.
Efficient Numerical Methods and Information-Processing Techniques for Modeling Hydro- and Environmental Systems
305 p. 2005 [3-540-24146-9]

Vol. 20: Zohdi T.I., Wriggers P.
Introduction to Computational Micromechanics
196 p. 2005 [3-540-22820-9]

Vol. 19: McCallen R., Browand F., Ross J. (Eds.)
The Aerodynamics of Heavy Vehicles: Trucks, Buses, and Trains
567 p. 2004 [3-540-22088-7]

Vol. 18: Leine, R.I., Nijmeijer, H.
Dynamics and Bifurcations of Non-Smooth Mechanical Systems
236 p. 2004 [3-540-21987-0]

Vol. 17: Hurtado, J.E.
Structural Reliability: Statistical Learning Perspectives
257 p. 2004 [3-540-21963-3]

Vol. 16: Kienzler R., Altenbach H., Ott I. (Eds.)
Theories of Plates and Shells: Critical Review and New Applications
238 p. 2004 [3-540-20997-2]

Vol. 15: Dyszlewicz, J.
Micropolar Theory of Elasticity
356 p. 2004 [3-540-41835-0]

Vol. 14: Frémond M., Maceri F. (Eds.)
Novel Approaches in Civil Engineering
400 p. 2003 [3-540-41836-9]

Vol. 13: Kolymbas D. (Eds.)
Advanced Mathematical and Computational Geomechanics
315 p. 2003 [3-540-40547-X]

Vol. 12: Wendland W., Efendiev M. (Eds.)
Analysis and Simulation of Multifield Problems
381 p. 2003 [3-540-00696-6]

Vol. 11: Hutter K., Kirchner N. (Eds.)
Dynamic Response of Granular and Porous Materials under Large and Catastrophic Deformations
426 p. 2003 [3-540-00849-7]

Vol. 10: Hutter K., Baaser H. (Eds.)
Deformation and Failure in Metallic Materials
409 p. 2003 [3-540-00848-9]

Vol. 9: Skrzypek J., Ganczarski A.W. (Eds.)
Anisotropic Behaviour of Damaged Materials
366 p. 2003 [3-540-00437-8]

Vol. 8: Kowalski, S.J.
Thermomechanics of Drying Processes
365 p. 2003 [3-540-00412-2]

Vol. 7: Shlyannikov, V.N.
Elastic-Plastic Mixed-Mode Fracture Criteria and Parameters
246 p. 2002 [3-540-44316-9]

Vol. 6: Popp K., Schiehlen W. (Eds.)
System Dynamics and Long-Term Behaviour of Railway Vehicles, Track and Subgrade
488 p. 2002 [3-540-43892-0]

Vol. 5: Duddeck, F.M.E.
Fourier BEM: Generalization of Boundary Element Method by Fourier Transform
181 p. 2002 [3-540-43138-1]

Vol. 4: Yuan, H.
Numerical Assessments of Cracks in Elastic-Plastic Materials
311 p. 2002 [3-540-43336-8]

Vol. 3: Sextro, W.
Dynamical Contact Problems with Friction: Models, Experiments and Applications
159 p. 2002 [3-540-43023-7]

Vol. 2: Schanz, M.
Wave Propagation in Viscoelastic and Poroelastic Continua
170 p. 2001 [3-540-41632-3]

Vol. 1: Glocker, C.
Set-Valued Force Laws: Dynamics of Non-Smooth Systems
222 p. 2001 [3-540-41436-3]

Uncertainty Assessment of Large Finite Element Systems

Christian A. Schenk
Gerhart I. Schuëller

Dr. Christian A. Schenk
TIWAG-Tiroler Wasserkraft AG
Eduard-Wallnöfer-Platz 2
6020 Innsbruck
Austria
christian.schenk@tiwag.at

Professor Dr.-Ing. habil. Gerhart I. Schuëller, Ph.D.
Leopold-Franzens University
Technikerstr. 13
6020 Innsbruck
Austria
mechanik@uibk.ac.at

With 61 Figures and 3 Tables

ISSN 1613-7736

ISBN-10 3-540-25343-2 **Springer Berlin Heidelberg New York**

ISBN-13 978-3-540-25343-3 **Springer Berlin Heidelberg New York**

Library of Congress Control Number: 2005925344

Springer is a part of Springer Science+Business Media
springeronline.com

Printed in Germany

Typesetting: Data conversion by the authors.
Final processing by PTP-Berlin Protago-TEX-Production GmbH, Germany
Cover-Design: design & production GmbH, Heidelberg
Printed on acid-free paper 89/3020/Yu - 5 4 3 2 1 0

Preface

The treatment of uncertainties in the analysis of engineering structures remains one of the premium challenges in modern structural mechanics. It is only in recent years that the developments in stochastic and deterministic computational mechanics began to be synchronized. To foster these developments, novel computational procedures for the uncertainty assessment of large finite element systems are presented in this monograph. The stochastic input is modeled by the so-called Karhunen-Loève expansion, which is formulated in this context both for scalar and vector stochastic processes as well as for random fields. Particularly for strongly non-linear structures and systems the direct Monte Carlo simulation technique has proven to be most advantageous as method of solution. The capabilities of the developed procedures are demonstrated by showing some practical applications.

Innsbruck,
March 2005

Christian A. Schenk
Gerhart I. Schuëller

Contents

1

Objectives

Complex deterministic systems under deterministic loading are generally analyzed by the finite element method. This is mainly due to the most versatile applicability of this method as well as the sophisticated solution algorithms available for purely deterministic analyses, see e.g. [37, 14, 66, 64, 16].
Quite frequently, however, neither the loading nor some structural parameters are deterministic quantities, i.e. they exhibit an inherent randomness and hence forcing also the response of the system to be an uncertain quantity. Deterministic analysis thus generally represents only an *approximation* of the actual reality, i.e. because of unavoidable uncertainties in the structural properties as well as in loading processes.
In many cases it suffices to take into account this inherent randomness by introducing empirically chosen safety factors, tacitly accepting all the disadvantages of such an approach, such as overly conservative designs, etc. In this regard it has to be mentioned, that in the past there was also – due to the limited computational facilities available – no other possibility of proceeding. Nowadays, with ever improving modern computational facilities at hand, one can certainly do better.
Since a stochastic analysis is significantly more involved than its deterministic counterpart, a large number of the procedures available in stochastic mechanics are applicable only to small systems, i.e. single or multi degree of freedom systems, respectively, or to structures, where analytical solutions are available. Consequently, the mathematical and/or mechanical model used in many of these investigations simplify reality in such a way that the usefulness of the numerical prediction as a whole might be questionable. Many approaches do not allow to incorporate statistical data of measurements into the analysis. For example, the assumption of homogeneity or stationarity of a stochastic process is most often an idealization and hence a restriction, yet this mathematical model rarely fits the data.
From the awareness of the random properties of the parameters in structural analysis the field of stochastic mechanics in the late fifties of the last century has emerged. See e.g. [124, 119] for a state-of-the-art review on

the developments and the current status of the fields of stochastic structural analysis and reliability in general and computational stochastic mechanics in particular. Still, in 2004, the analysis of large finite element systems involving uncertain quantities is subject of intensive research efforts, see e.g. [67, 118, 120, 125, 126]. The information captured by the stochastic response, either in terms of statistical moments or in probability densities, is considerably more comprehensive than that of the deterministic response – which in fact describes what one could compare to the mean function only.
The acceptance of the notion of a stochastic analysis in engineering practice is strongly related to the state-of-the-art of deterministic analysis as used within the stochastic analysis, hence implying that stochastic analysis is in competition with deterministic analysis. It thus seems to be advantageous, to incorporate deterministic algorithms and approaches into a stochastic analysis wherever this is feasible. This ensures on one hand, that current developments in deterministic mechanics can be directly exploited and on the other hand, that engineers familiar with deterministic mechanics might feel more comfortable with stochastic approaches.

These introductory remarks allow to define the objectives of this monograph, which are in the description of the computational procedures allowing the performance of an uncertainty analysis of large finite element systems. Special attention is devoted to the applicability of the developed procedures in the engineering practice, i.e. the procedures should

(i) be capable to handle large finite element systems,
(ii) rely on a high degree on deterministic solution algorithms in order to allow the to use of commercial finite element packages,
(iii) be able to incorporate measured data if available.

Part II of this monograph encompasses the theoretical background of the methods subsequently used for practical applications as discussed in Part III. Some of these procedures are well known and hence summarized only briefly. The rational treatment of uncertainties forms the basis of the developments as outlined in this monograph. For this purpose the notions of random variables, processes and fields, respectively are described and discussed. Their Gaussian properties play a central role in the theory as well as in the application of stochastic methods. The most common representations of stochastic processes are spectral representation, Karhunen-Loève and polynomial chaos representation, respectively, as well as wavelets representation, where for practical applications the Karhunen-Loève expansion [73] of the covariance function proves to be most efficient. Since it is applicable for Gaussian processes only, uncertainties in terms of second statistical moments (variances) can be quantified easily, even for larger structural systems.The polynomial chaos expansion can be viewed as a generalization of the Karhunen-Loève expansion and is also applicable for non-Gaussian processes and fields, respectively.
The most generally applicable tool to compute the response of stochastically

excited structures – with either deterministic or stochastic properties respectively – is the Monte Carlo simulation procedure, see e.g. [45]. It is based on random sampling, where the laws of statistics are exploited to derive information on the variability of the response. For calculating the confidence in the calculated response, statistical estimation procedures are applied. This also governs the number of simulations required to reach a certain accuracy of the estimator.
For non linear systems the method of statistical equivalent linearization is applied quite frequently. While the Monte Carlo simulation procedure provides the information of the entire probability structure, this method provides only the second order statistics of the response, see e.g. [102, 109].
For larger non linear finite element systems, maintaining the superposition principle that is applicable for linear systems is of paramount importance, particularly for computational efficiency. Again, the basic idea for the computation of the stochastic response is based on the Karhunen-Loève expansion of the stochastic loading. This is accomplished by computing for each Karhunen-Loève vector of the loading the corresponding Karhunen-Loève vector of the response. Since not all second moment characteristics of all degrees of freedom are of equal importance, the computational efforts can be reduced significantly, i.e. by reduction of the dimension of the problem.

In order to place the procedures described in this monograph into perspectives, historical and also other current developments are illuminated below. In this context, two areas of particular practical interest are treated in detail, i.e. the stability analysis of systems with random imperfections and the dynamic analysis of deterministic systems subjected to stochastic loading, respectively.
The stability problem of thin shell structures has received worldwide particular attention because of the distinct discrepancy between experiments and classical numerical prediction of the buckling loads. The combined effects of non-linearity and imperfections, respectively, are the reason for this discrepancy and, moreover, for the large scatter observed in experimentally determined limit loads [15, 61].
So far most of the research work carried out in this field concentrates on the effects of traditional, i.e. geometric imperfections. However, other sources of imperfections may have also a strong effect on the critical load, such as thickness imperfections, non-perfect boundary conditions, misalignment in the loading, fluctuations in material properties, etc. In this context – and already at an early stage – measurements of geometric imperfections revealed that even shells from the same manufacturing process differ in shape and magnitude. It is this inherent randomness in geometric and also other imperfections, introduced during the production and construction process, that reduces the evidence of deterministic predictions of the critical load in particular of cylindrical shells considerably. Therefore, the design of thin shell structures at present based on either deterministic classical, asymptotic or

general non-linear analysis, respectively, still needs to be corrected on the basis of experimental data by using empirical, so-called "knock down" factors, leading to potentially overly conservative designs. This lower bound concept may be justified in cases where the total weight of the shell structure as well as the cost of the material do not play a significant role in the design process. In all other cases, a more general stability analysis should be applied taking into account the stochastic nature of the stability behavior of shells.
Nowadays highly sophisticated numerical methods are available, which are capable to incorporate several sources of imperfections into the analysis. However, these imperfections are described by means of deterministic parameters, allowing not to predict the aforementioned scatter in the critical load observed in experiments. Under these circumstances it seems to be quite natural to introduce a concept for a stability analysis, which allows to take into account the uncertainties in order to be able to predict the complex stability behavior of thin shell structures. It should be noted, that such an analysis should rest on the same sophisticated mechanical and mathematical formulations in structural stability as its purely deterministic counterpart (e.g. in terms of worst case studies) and thus yielding more realistic results.

A probabilistic approach in structural stability for imperfection sensitive structures dates already back to the late fifties (see e.g. [19]), followed by several other investigators, e.g. [46], [2], [110] and [58]). However, due to the lack of experimental evidence about the type of geometric imperfections that occur in practice, the aforementioned investigators had to work with some form of idealized, i.e assumed, imperfection distribution. A numerical measure for the significance of the various possible shapes of imperfections on the buckling strength of a structure in terms of reliability analysis is suggested in [87]. The Fourier coefficients of measured imperfections are treated in [42] as random variables, whereby the histogram of the limit load has been obtained either by Monte Carlo simulation or approximate methods. Common to these approaches, see also [7, 28, 159, 137], is the determination of the critical load by means of analytical or semi-analytical procedures. This may also be the reason for the stochastic representation of geometric imperfections by means of a two dimensional Fourier series with random Fourier coefficients, since an analytical buckling analysis of e.g cylindrical shells yields a two dimensional Fourier series representation of the critical modes.
It is well known, that severe shortcomings of analytical or semi-analytical approaches, respectively, for predicting buckling, are on one hand their limited capability in modeling more complex shell structures and on the other hand the implementation of a more general type of imperfection. For example, many practical applications of cylindrical shells do require the presence of either reinforced or unreinforced large cutouts, which are in most cases not amenable for analytical solutions (see e.g. [1, 144, 156, 127, 164, 60, 145, 149, 3]) and require in general a non-linear analysis. Moreover, linear bifurcation theory can not take into account significant stress redistributions taking place

near the edges of the cutouts, leading possibly to overly conservative results. It is thus clear, that more modern approaches which consider random imperfections, the finite element method is the method of first choice, see e.g. [18, 31, 32, 83, 71, 138, 17, 8, 146], where some idealized second moment properties of imperfections have been assumed in most cases. Considerable research has been done in the field of axially compressed cylindrical shells with random traditional, i.e. initial geometric imperfections, but to the authors knowledge no studies on the individual and combined effects of random boundary imperfections and random geometric imperfections on the buckling behavior of axially compressed cylindrical shells exist. As pointed out in [4], the sensitivity of anisotropic shells to boundary imperfections may exceed significantly that of traditional geometric imperfections, implying that boundary imperfections do have a crucial influence on the value as well as on the scatter of the limit load. Several studies have been presented dealing with the effect of a in thin shell structures, see e.g. [59]. Again, to the authors' knowledge, the effect of random geometric imperfections on the critical load of axially compressed cylindrical shells with a cutout has not been investigated so far. This may partially be due to the fact, that – depending on the size of the cutout – the cutout itself represents a predictable imperfection that might be more significant than unavoidable geometric imperfections. In this case, of course, there is no need to incorporate imperfections into the analysis. However, if the cutout is small, e.g. the effect of the imperfection introduced by the cutout lies in the range of that of geometric imperfections, also geometric imperfections should be incorporated into the analysis.

The various methods available to analyze large deterministic linear finite element systems subjected to Gaussian white or linearly filtered Gaussian white noise, respectively, can be distinguished into methods operating in the time or in frequency domain, respectively, see e.g [36, 72, 141, 75]. The response in this case is also Gaussian and these methods rationally quantify the uncertainties in structural analysis in terms of statistical second moment characteristics, i.e. the mean, covariance and/or the power spectral density function, respectively. For stationary solutions of linear structures, the spectral density of the response can be computed from the spectral density of the excitation using frequency response functions in the frequency domain. Solutions for non-stationary excitation, particularly in the frequency domain, are rather involved and even intractable for larger finite element systems. In this case, quite frequently, methods for the solution of the so-called Lyapunov matrix differential equation in the time domain (e.g. [53]), are applied. Whether or not systems are proportionally damped, the solution of the Lyapunov equation can be transformed into a reduced subspace using either modal analysis or complex modal analysis, respectively, thus being able to calculate the stochastic response of large finite element structures efficiently. However, in these cases, the white noise input process is quite often filtered and/or modulated in time, see e.g. [163, 81] for a more sophisticated model of earthquake excitation. Excitation models based on white noise do work well for single and multi degree of freedom systems, see e.g. [89, 10], but are not adequate for finite element models with thousands of degrees of freedom, because of unrealistic high frequency excitations and responses, respectively. Moreover, white noise always represents a mathematical abstraction of a real excitation process and hence may not easily adjusted to second moment properties of potentially available statistical data. Alternatively, the excitation can be formulated by the Karhunen-Loève representation (e.g. [39, 73, 49, 98, 139]). For a finite number of terms, the associated Karhunen-Loève representation of the response can be determined quite easily by deterministic analysis and the frequency range can be controlled in a simple manner.

The situation is quite different for non-linear systems subjected to additive and/or multiplicative Gaussian white noise, where the response forms also a vector Markov process. It has been revealed by a benchmark study on non-linear stochastic structural dynamics [148, 123], the procedures available to analyze such systems are limited – with only a few exceptions – to multi degree of freedom systems, see e.g. [96]. Methods for calculating higher order moments or response probabilities like the so-called moment equations or Fokker-Planck equations increase in complexity at least exponentially with the state space dimension. Hence so far they are not applicable to larger systems. Approaches based on the numerical solution of the Fokker-Planck equation using Galerkin, finite element and path integral methods, respectively, are feasible only for state space dimensions up to four, see e.g. [143, 82, 63, 155, 158] . The exceptions mentioned above which are capable to analyze the stochastic response of large systems are Monte Carlo simulation, see e.g. [152, 56] and equivalent statistical linearization, see e.g. [102]. Reviewing the literature with respect to equivalent statistical linearization, it is apparent that procedures for general non-linear elements have not been developed and tested to such a degree that this linearization technique can be recommended for general use. Research has been devoted almost exclusively to one-dimensional elements, where the so-called "Bouc-Wen" hysteretic model, see e.g. [12], has been discussed and studied most extensively.
A multi-dimensional non-linear element requires a numerical integration in the dimension according to the length of the state vector describing the element - obviously this is not feasible for higher state space dimensions. Two basic approaches are available when applying equivalent statistical linearization: Complex modal analysis and stochastic versions of direct step by step integration schemes. The use of complex modal analysis has been suggested e.g. in [29] for treating non-stationary filtered white noise excitation for earthquake problems. Due to the increase in dimension of the system equations and emerging asymmetric matrices, complex modal analysis is computationally considerably more involved than classical modal analysis. Hence, complex modal analysis is not well suited for treating larger finite element systems. Updating procedures, as suggested in [97], based on the solution of the classical eigenvalue

problem and component mode synthesis overcome the aforementioned shortcomings of complex modal analysis, thus making the analysis of non-linear finite element systems feasible. Still, this approach cannot be recommended for non-zero mean problems, where the modal properties change fundamentally in each time step. Stochastic versions of special integration schemes have been suggested e.g. in [151, 79, 86]. All these proposed procedures operate explicitly on the covariance matrix and hence – due to storage requirements – cannot be applied efficiently to finite element systems, particularly for systems with a large number of degrees of freedom.
Recently a solution technique which avoids the storage and direct integration of the full covariance matrix has been developed [98, 100]. There it is suggested that the covariance matrix of the response is represented by a truncated Karhunen-Loève expansion. Continuous white noise is approximated by discrete noise applied at one instant in each time step. The Karhunen-Loève vectors can be integrated by any available and appropriate deterministic step by step integration scheme. Due to the discrete white noise loading, the Karhunen-Loève vectors have to be updated in each time step, rendering the calculation of the stochastic response rather involved.

2 Outline

The monograph is structured as follows:

Part I including Chapts. 3–6 reviews briefly the theoretical background of the deterministic methods and procedures as used in subsequent chapters. From the proposed procedures, particularly from the mathematical point of view, the so-called spectral theorem is most important. It is particularly useful when dealing with finite dimensional operators.

Part II comprises probabilistic methods and procedures relevant for the practical applications shown in Part III. Starting with a general chapter on the rational treatment of uncertainties in computational stochastic mechanics (Chap. 7), the focus of Chapt. 8 lies on the discrete version of the Karhunen-Loève expansion. Well known tools in computational stochastic mechanics such as the Monte Carlo simulation technique and equivalent statistical linearization are treated in Chapts. 9 and 10. Chapter 11 describes an efficient procedure for the calculation of statistical second moment characteristics of large linear and non-linear finite element systems subjected to stochastic dynamic loading.

Some practical applications of the procedures treated in Part I and II are shown in Part III. Chapter 12 concentrates on the uncertainty assessment of large non-linear imperfection sensitive finite element systems, special emphasis is given to the stability analysis of axially compressed cylindrical shells. Chapt. 13 is devoted to random vibration problems, in particular to the calculation of the response of a large 6-story office building subjected to dynamic stochastic loading.

The monograph is concluded with Part IV. Appendix A describes briefly an imperfection database, which provided the basis for the estimation of the second moment properties of geometric and boundary imperfections, respectively, of cylindrical shells. For the sake of completeness, the fundamentals of the Lyapunov matrix differential equation are shown in Appendix B.

Part I

Deterministic Methods and Procedures

3

Spectral Analysis of Finite Dimensional Operators

A great number of engineering problems can be viewed as an input to output mapping, as shown in Fig. 3.1, where the system represents the law for the mapping between input and output. Mathematically speaking, the system represents an operator, which transforms input to output. The systems treated in this monograph can be described most generally in the form

$$L\boldsymbol{u}(t) = \boldsymbol{f}(t) , \tag{3.1}$$

where L denotes the operator, $\boldsymbol{u}(t)$ denotes the displacement vector and $\boldsymbol{f}(t)$ is an external force vector. In particular, the operator for dynamical non-linear finite element systems can be written, denoting $\boldsymbol{M}$ as the mass matrix, according to

$$L_D\boldsymbol{u}(t) = \boldsymbol{M}\frac{\mathrm{d}^2\boldsymbol{u}(t)}{\mathrm{d}t^2} + \boldsymbol{r}\Big(\boldsymbol{u}(t), \frac{\mathrm{d}\boldsymbol{u}(t)}{\mathrm{d}t}\Big) = \boldsymbol{M}\ddot{\boldsymbol{u}}(t) + \boldsymbol{r}(\boldsymbol{u}(t), \dot{\boldsymbol{u}}(t)) , \tag{3.2}$$

while for systems where inertia and damping forces can be omitted, L reads

$$L_S\boldsymbol{u}(t) = \boldsymbol{r}(\boldsymbol{u}(t)) . \tag{3.3}$$

In both equations (3.2) and (3.3), $\boldsymbol{r}(\cdot)$ denotes a non-linear restoring force vector which is governed by the state of the system. It is known from experimental data that loading conditions, geometry, material properties, etc. exhibit inherent fluctuations, that can generally not not be described by deterministic means. Fluctuations or variations in the system input cause also fluctuations in the system output. In this monograph, the external force vector $\boldsymbol{f}(t)$ in (3.2) and the initial state of the displacement vector $\boldsymbol{u}(0)$ in (3.3) are considered to be uncertain, i.e. these quantities can be represented by stochastic processes. The Karhunen-Loève expansion, see Chap. 8, will be used in order to model the second moment statistics of $\boldsymbol{f}(t)$ in (3.2) and $\boldsymbol{u}(0)$ in (3.3), so that the integral operator L_K defined by

$$L_K\psi(s) = \int_a^b \Gamma(t,s)\psi(s)\mathrm{d}s , \tag{3.4}$$

where $\Gamma(t,s)$ denotes the covariance kernel of the stochastic input and $\psi(s)$ a corresponding eigenfunction, is of paramount importance for the subsequent chapters. Although (3.2) and (3.3) describe non-linear operators, the solution

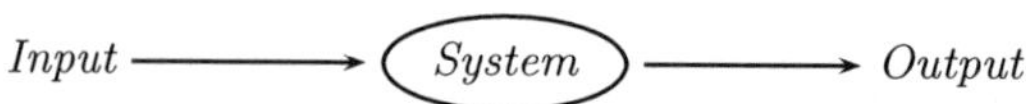

Fig. 3.1. Schematic drawing of a general mechanical system.

of these equations will be obtained by some type of piecewise linearization. Keeping this in mind, equations (3.2)–(3.4), i.e. matrix differential (L_D), matrix (L_S) and integral equations (L_K), respectively, can be treated in a unified manner by linear operator theory, see e.g. [84]. In this regard, it is worth mentioning that in computational stochastic mechanics, due to necessary discretization, equations (3.2)-(3.4) are always defined over finite dimensional domains.

In this chapter, a brief overview is given of the theoretical background needed for the development of computational procedures in order to solve (3.2) and (3.3) in context with (3.4). Hereby, the so-called spectral theorem of linear, finite dimensional, symmetric operators plays a central role. As already mentioned, the spaces over which (3.2)-(3.4) are defined are - due to spatial and/or temporal discretization - always finite dimensional. As a consequence, one can leave aside topological considerations (Cauchy sequence, completeness, etc), i.e. one has not to resort on the theory of infinite dimensional Hilbert spaces. This would be necessary, for example, if distributed parameter systems or continuous stochastic processes are considered. The reason why finite dimensional vector spaces are much more easier to deal with is based on the following statements [84]:

(i) all finite dimensional normed linear spaces are Hilbert spaces,
(ii) every linear subspace of a finite dimensional normed linear space is closed,
(iii) boundedness implies total boundedness, and
(iv) all linear transformations are continuous.

A most important consequence of these statements is that convergence of an orthogonal series in such finite dimensional spaces is not an issue. It is well known, that finite dimensional linear transformations can be represented by matrices, a representation of linear operators which is extensively used in computational mechanics. A linear transformation and its matrix representation, however, are not the same. While a linear transformation defines a rule for assigning the input to the output, the corresponding matrix represents the transformation relative to the bases of the input and output.

Let L be a linear symmetric operator defined on a finite dimensional vector space $\mathcal{V} = \mathcal{R}^n$. Basically, the idea of the spectral theorem is to break up $\mathcal{V}$ into a finite number of parts in such a way, that the operation of L on each part is particularly simple. In this context, the definition of orthogonal projections P_i appears to be useful. Orthogonal projections satisfy the following conditions: a) P_i is linear and b) $P_i^2 = P_i$ (idempotent), see e.g. [57]. The so-called resolution of the identity, $\{P_i\}_{i=1}^n$ satisfies the condition

$$\sum_{i=1}^{n} P_i = 1 \, . \tag{3.5}$$

This allows to define the spectral theorem. Let $\{P_i\}_{i=1}^n$ be a resolution of the identity on a finite dimensional normed linear vector space and let

$$L = \sum_{i=1}^{n} \lambda_i P_i \, , \tag{3.6}$$

where $\{\lambda_i\}_{i=1}^n$ are distinct scalars. Then there exists an orthonormal basis $\{\phi_i\}_{i=1}^n$ of eigenvectors of L, defined by

$$L\phi_i = \lambda_i \phi_i \, . \tag{3.7}$$

Moreover, every vector $x \in \mathcal{V}$ can be written in the form (using $(\cdot)$ for the inner product)

$$x = \sum_{i=1}^{n} (x, \phi_i)\phi_i \tag{3.8}$$

and

$$Lx = \sum_{i=1}^{n} \lambda_i (x, \phi_i)\phi_i \, . \tag{3.9}$$

Equation (3.8) represents the well known expansion theorem. By comparing (3.5) with (3.9) one obtains

$$P_i = \phi_i \phi_i^T \tag{3.10}$$

and it follows that $P_i = P_i^T$. Especially useful when dealing with matrices, the spectral theorem can be formulated in a different manner. Let $\mathcal{V}$ be a finite dimensional vector space and $L : \mathcal{V} \rightarrow \mathcal{V}$ be a symmetric linear operator. Then there is a basis $\{\phi\}_{i=1}^n$ in $\mathcal{V}$, that the matrix $\boldsymbol{\Lambda}$ that represents $\mathcal{V}$ in this basis is a diagonal matrix, i.e.

$$\boldsymbol{\Lambda} = \begin{bmatrix} \lambda_1 & 0 & \dots & 0 \\ 0 & \lambda_2 & \ddots & \vdots \\ \vdots & \ddots & \ddots & 0 \\ 0 & \dots & 0 & \lambda_n \end{bmatrix} \, . \tag{3.11}$$

It follows from above, that the spectral theorem for a linear operator L with respect to its matrix representation $\boldsymbol{L}$ represents an eigenvalue decomposition of $\boldsymbol{L}$, i.e.

$$\boldsymbol{L}\boldsymbol{\phi}_i = \lambda_i \boldsymbol{\phi}_i \,. \tag{3.12}$$

Some important properties of the eigenvalue decomposition of real, symmetric matrices are, see e.g. [78]:

(i) all eigenvectors $\boldsymbol{\phi}$ and eigenvalues λ are real,
(ii) two eigenvectors corresponding to distinct eigenvalues are orthogonal,
(iii) eigenvalue λ_i with algebraic multiplicity m_i does have exactly m_i linearly independent corresponding eigenvectors (not uniquely defined), i.e. $\{\boldsymbol{\phi}\}_i^{m_i}$, which can be rendered to be orthogonal (algebraic multiplicity is equal geometric multiplicity), thus
(iv) $\boldsymbol{L}$ is non-defective (i.e. algebraic multiplicity does not exceed geometric multiplicity),
(v) $\boldsymbol{L}$ is diagonalizable by means of a similarity transformation of its modal matrix $\boldsymbol{\Phi}$, i.e. $\boldsymbol{\Phi}^T\boldsymbol{L}\boldsymbol{\Phi} = \boldsymbol{\Lambda}$, where the eigenvalue matrix $\boldsymbol{\Lambda}$ is defined according to (3.11) and

$$\boldsymbol{\Phi} = \left[\boldsymbol{\phi}_1 \; \boldsymbol{\phi}_2 \; \cdots \; \boldsymbol{\phi}_n\right] \,. \tag{3.13}$$

While the eigenvalue problem of a single matrix is often referred to the standard eigenvalue problem and is needed e.g. in context with a linearized buckling analysis (see Sec. 5.4), the eigenvalue problem of a real symmetric matrix $\boldsymbol{K}$ and a real symmetric positive definite matrix $\boldsymbol{M}$ given by

$$\boldsymbol{K}\boldsymbol{\phi}_i = \lambda_i \boldsymbol{M}\boldsymbol{\phi}_i \,, \tag{3.14}$$

is denoted as the generalized eigenvalue problem, which occurs in the dynamic analysis of finite element systems in context with mode superposition methods. In principal, the generalized eigenvalue problem can be reduced to the standard eigenvalue problem by a Cholesky decomposition of $\boldsymbol{M}$, i.e.

$$\boldsymbol{M} = \boldsymbol{R}^T\boldsymbol{R} \,. \tag{3.15}$$

By substituting (3.15) into (3.14) one obtains the standard eigenvalue problem

$$\tilde{\boldsymbol{K}}\tilde{\boldsymbol{\phi}}_i = \lambda_i \tilde{\boldsymbol{\phi}}_i \,, \tag{3.16}$$

where

$$\tilde{\boldsymbol{\phi}}_i = \boldsymbol{R}\boldsymbol{\phi}_i \,, \qquad \tilde{\boldsymbol{K}} = (\boldsymbol{R}^{-1})^T\boldsymbol{K}\boldsymbol{R}^{-1} \,. \tag{3.17}$$

It should be mentioned that if the eigenvalues $\tilde{\boldsymbol{\phi}}_i$ of $\tilde{\boldsymbol{K}}$ are scaled such that they are mutually orthonormal, i.e.

$$\tilde{\boldsymbol{\phi}}_i\tilde{\boldsymbol{\phi}}_j = \delta_{ij} \,, \tag{3.18}$$

where δ_{ij} is the Kronecker delta, then the eigenvectors $\boldsymbol{\phi}$ associated with (3.14) are normalized such that

$$\boldsymbol{\phi}_i^T \boldsymbol{M} \boldsymbol{\phi}_j = \delta_{ij}, \qquad \boldsymbol{\phi}_i^T \boldsymbol{K} \boldsymbol{\phi}_j = \lambda_i \tag{3.19}$$

holds.

4

Finite Element Method

In this chapter, the fundamentals of the displacement-based finite element method are reiterated in a concise form, details can be found in a number of textbooks and the references therein. For larger finite element systems, model reduction techniques are commonly applied in order to save computational efforts, so that in this chapter two most commonly applied model reduction techniques – the mode displacement method and the mode acceleration method – are reviewed. Various formulations of the finite element method have been developed in the past years, e.g. the displacement-based finite element formulation and several mixed finite element formulations. However, the most frequently used formulation of the finite element method is the displacement-based formulation, i.e. the displacements are the unknown variables which have to satisfy certain boundary conditions. Once the displacements have been determined, strains and stresses can be calculated. The displacement-based finite element analysis can be derived by the principal of virtual displacements (or work), which in turn is equivalent to setting the first variation of the total potential energy equal to zero. The principal of virtual work can be cast into the form, see e.g. [14],

$$\underbrace{\int_V \bar{\boldsymbol{\epsilon}}^T \boldsymbol{\tau}\, dV}_{\text{internal virtual work}} = \underbrace{\int_V \bar{\boldsymbol{u}}^T \boldsymbol{f}_B\, dV + \int_{S_f} \bar{\boldsymbol{u}}_S^T \boldsymbol{f}_S\, dS + \sum_i \bar{\boldsymbol{u}}^{iT} \boldsymbol{f}_C^i}_{\text{external virtual work}} , \tag{4.1}$$

implying that the total internal virtual work is equal to the total external virtual work. In (4.1) the virtual displacements are denoted by $\bar{\boldsymbol{u}}$, $\bar{\boldsymbol{\epsilon}}$ denote the corresponding virtual strains and $\boldsymbol{f}_B$, $\boldsymbol{f}_C$ and $\boldsymbol{f}_{S_f}$ are the externally applied body forces (per unit volume), surface tractions (per unit surface area) and concentrated forces, respectively. For a structure with given geometry, applied forces, boundary conditions, material stress-strain law and initial stresses, respectively, the displacements $\boldsymbol{u}$ and associated strains $\boldsymbol{\epsilon}$ and $\boldsymbol{\tau}$ can be calculated. The vector $\boldsymbol{\tau}$ contains the correct stresses if and only if (4.1) is satisfied for any arbitrary virtual displacement field $\bar{\boldsymbol{u}}$ that is continuous and zero at and corresponding to the prescribed displacements on the surface [14].

The displacements $\boldsymbol{u}^{(e)}$ of each element in element specific (local) coordinates are related to all displacements $\boldsymbol{u}$ in global coordinates by the so-called displacement interpolation matrix $\boldsymbol{H}^{(e)}$, i.e.

$$\boldsymbol{u}^{(e)} = \boldsymbol{H}^{(e)}\boldsymbol{u} \,. \tag{4.2}$$

In a similar way, the element strains $\boldsymbol{\epsilon}^{(e)}$ can be calculated

$$\boldsymbol{\epsilon}^{(e)} = \boldsymbol{B}^{(e)}\boldsymbol{u} \,, \tag{4.3}$$

where $\boldsymbol{B}^{(e)}$ is the strain-displacement matrix. The element stresses $\boldsymbol{\tau}^{(e)}$ are related to the element strains and the initial stresses $\boldsymbol{\tau}^{I(e)}$ according to

$$\boldsymbol{\tau}^{(e)} = \boldsymbol{C}^{(e)}\boldsymbol{\epsilon}^{(e)} + \boldsymbol{\tau}^{I(e)} \,, \tag{4.4}$$

where the material law for each element is specified in $\boldsymbol{C}^{(e)}$. Assuming unit virtual displacements in (4.1), together with (4.2)–(4.4) one arrives, e.g. for undamped dynamical systems, at the linear equation of motion

$$\boldsymbol{M}\ddot{\boldsymbol{u}} + \boldsymbol{K}\boldsymbol{u} = \boldsymbol{f} \,, \tag{4.5}$$

where the mass and the stiffness matrix is given by

$$\boldsymbol{M} = \sum_e \int_{V^{(e)}} \rho^{(e)} \boldsymbol{H}^{(e)T} \boldsymbol{H}^{(e)} dV^{(e)} \tag{4.6}$$

and

$$\boldsymbol{K} = \sum_e \int_{V^{(e)}} \boldsymbol{B}^{(e)T} \boldsymbol{C}^{(e)} \boldsymbol{B}^{(e)} dV^{(e)} \,, \tag{4.7}$$

respectively. The load vector is specified by

$$\boldsymbol{f} = \boldsymbol{f}_B + \boldsymbol{f}_C - \boldsymbol{f}_I + \boldsymbol{f}_C \,, \tag{4.8}$$

where the components of $\boldsymbol{f}$ are defined by

$$\boldsymbol{f}_B = \sum_e \int_{V^{(e)}} \boldsymbol{H}^{(e)T} \boldsymbol{f}^{B(e)} dV^{(e)} \,, \tag{4.9}$$

$$\boldsymbol{f}_S = \sum_e \int_{S_1^{(e)},\dots,S_q^{(e)}} \boldsymbol{H}^{S(e)T} \boldsymbol{f}^{S(e)} dS^{(e)} \tag{4.10}$$

and

$$\boldsymbol{f}_I = \sum_e \int_{V^{(e)}} \boldsymbol{B}^{(e)T} \boldsymbol{\tau}^{I(e)} dV^{(e)} \,. \tag{4.11}$$

The approximation in the finite element method lies in the approximation of the differential equilibrium, while the nodal point equilibrium and the element equilibrium are – independent from the mesh used – always satisfied. Thus for the nodal point equilibrium the following relation holds

$$\boldsymbol{r} = \sum_e \int_{V^{(e)}} \boldsymbol{B}^{(e)T} \boldsymbol{\tau}^{(e)} dV^{(e)} \equiv \boldsymbol{f} \ , \tag{4.12}$$

where $\boldsymbol{r}$ denote the element nodal point forces. This relation represents, most generally, the starting point of a non-linear finite element analysis.

5

Non-Linear Static Analysis

The state of the art in the capability to predict the stability behavior of shell structures refers to areas such as asymptotic analysis [69] and general non-linear analysis (see e.g. [16, 30]). In both approaches, the solution of equilibrium equations under variation of a number of parameters such as load level, imperfection magnitude etc., is necessary. Of paramount interest is the determination of the so-called critical points. Two fundamental problems have to be distinguished in the field of structural stability: the buckling problem and the collapse or snap-through problem, respectively. The corresponding critical points are denoted as bifurcation points and limit points. By definition a bifurcation point describes the equilibrium state of a structure where distinct non-interacting solutions are possible. The limit point on the other hand characterizes the local maximum of the load-deflection curve. Contrary to perfect structures, imperfect structures do not show a clear separation of response modes. Instead, the response is a combination of all excited modes. Depending on the magnitude of the imperfection, it is possible that bifurcation points may vanish.

Asymptotic analysis rests on perturbation theory by expanding the solution into a power series, being asymptotically exact at the bifurcation point. While asymptotic analysis has contributed significant physical insight into the buckling process for a certain class of perfect as well as for imperfect shell structures, the general non-linear analysis allows to determine the equilibrium path of arbitrary structures. A general non-linear analysis – belonging to the class of continuation methods – allows to determine the response of a structure that is not confined to a particular region of the solution space. One of the key differences between the asymptotic analysis, frequently denoted as Koiter's method, and the general non-linear analysis is the use of iteration methods in the latter one, see e.g. [106]. Early applications of continuation methods were quite limited because of severe convergence difficulties in the neighborhood of limit points as well as their low computational efficiency due to necessary iterations. However, introducing adaptive continuation parameters accompanied by the fast evolution of digital computers, a general non-linear analysis is nowadays the most versatile technique for the investigation of the stability behavior of shell structures.

In a non-linear analysis, the equilibrium of the structure between the applied load vector $\boldsymbol{f}$ and the element nodal point forces $\boldsymbol{r}(\boldsymbol{u})$

$$ {}^{t}\boldsymbol{f} - {}^{t}\boldsymbol{r} = \boldsymbol{0} \,, \tag{5.1} $$

see (4.12), has to be determined. In (5.1) the superscript t describes the state of the system, rather than the time as used frequently in a dynamic analysis. Assuming that the equilibrium at state t is satisfied, the equilibrium for the state $t + \Delta t$

$$ {}^{t+\Delta t}\boldsymbol{f} - {}^{t+\Delta t}\boldsymbol{r} = \boldsymbol{0} \tag{5.2} $$

is sought. For the element nodal point forces at state $t + \Delta t$ one can write

$$ {}^{t+\Delta t}\boldsymbol{r} = {}^{t}\boldsymbol{r} + \Delta\boldsymbol{r} \,, \tag{5.3} $$

where $\Delta\boldsymbol{r}$ denotes the increment in nodal point forces associated with the increment $\Delta\boldsymbol{u}$ of the displacements from t to $t + \Delta t$. In general, the non-linear function $\boldsymbol{r}(\boldsymbol{u})$ can be expanded in a first order Taylor series around the state ${}^{t}\boldsymbol{u}$, i.e.

$$ \boldsymbol{r}(\boldsymbol{u}) \approx \boldsymbol{r}({}^{t}\boldsymbol{u}) + \left.\frac{\partial \boldsymbol{r}}{\partial \boldsymbol{u}}\right|_{{}^{t}\boldsymbol{u}} (\boldsymbol{u} - {}^{t}\boldsymbol{u}) \,, \tag{5.4} $$

where the Jacobian matrix is frequently denoted as the tangent stiffness matrix

$$ {}^{t}\boldsymbol{K} = \left.\frac{\partial \boldsymbol{r}}{\partial \boldsymbol{u}}\right|_{{}^{t}\boldsymbol{u}} . \tag{5.5} $$

Now the increment in nodal point forces can be approximated by

$$ \Delta\boldsymbol{r} = {}^{t}\boldsymbol{K} \Delta\boldsymbol{u} \,. \tag{5.6} $$

If ${}^{t+\Delta t}\boldsymbol{f}$ is independent of the displacement field ${}^{t+\Delta t}\boldsymbol{u}$, i.e. no so-called follower forces, then the displacement increment can be calculated by substituting (5.3) and (5.6) into (5.2) yielding

$$ {}^{t}\boldsymbol{K} \Delta\boldsymbol{u} = {}^{t+\Delta t}\boldsymbol{f} - {}^{t}\boldsymbol{r} \,, \tag{5.7} $$

thus one obtains the displacements at $t + \Delta t$ according to

$$ {}^{t+\Delta t}\boldsymbol{u} = {}^{t}\boldsymbol{u} + \Delta\boldsymbol{u} \,. \tag{5.8} $$

Because (5.8) is just an approximation of the actual displacements at $t + \Delta t$ due to (5.4), iterations are necessary in order to obtain the displacement field ${}^{t+\Delta t}\boldsymbol{u}$ within a given tolerance. The most frequently applied iteration schemes in a non-linear finite element analysis are the full Newton-Raphson scheme and the modified Newton-Raphson scheme.

5.1 Full Newton-Raphson Scheme

A characteristic of the full Newton-Raphson scheme is that the tangent stiffness matrix (5.5) is updated in every iteration s. Thus (5.2) is solved by the relations

$$ {}^{t+\Delta t}\boldsymbol{K}^{(s-1)}\Delta\boldsymbol{u}^{(s)} = {}^{t+\Delta t}\boldsymbol{f} - {}^{t+\Delta t}\boldsymbol{r}^{(s-1)} \tag{5.9} $$

and

$$ {}^{t+\Delta t}\boldsymbol{u}^{(s)} = {}^{t+\Delta t}\boldsymbol{u}^{(s-1)} + \Delta\boldsymbol{u}^{(s)} , \tag{5.10} $$

where

$$ {}^{t+\Delta t}\boldsymbol{K}^{(s-1)} = \left.\frac{\partial \boldsymbol{r}}{\partial \boldsymbol{u}}\right|_{{}^{t+\Delta t}\boldsymbol{u}^{(s-1)}} . \tag{5.11} $$

The initial condition for each iteration cycle are given by

$$ {}^{t+\Delta t}\boldsymbol{u}^{(0)} = {}^{t}\boldsymbol{u} , \qquad {}^{t+\Delta t}\boldsymbol{K}^{(0)} = {}^{t}\boldsymbol{K} , \qquad {}^{t+\Delta t}\boldsymbol{r}^{(0)} = {}^{t}\boldsymbol{r} . \tag{5.12} $$

5.2 Modified Newton-Raphson Scheme

In the full or true Newton-Raphson scheme the major computational time is spent for updating the tangent stiffness matrix (5.11) in every iteration. In the modified Newton-Raphson scheme, the tangent stiffness matrix is held constant as long as no convergence difficulties occur, thus (5.9) is replaced by

$$ {}^{\tau}\boldsymbol{K}\Delta\boldsymbol{u}^{(s)} = {}^{t+\Delta t}\boldsymbol{f} - {}^{t+\Delta t}\boldsymbol{r}^{(s-1)} , \tag{5.13} $$

where ${}^{\tau}\boldsymbol{K}$ denotes a tangent stiffness matrix corresponding to a previous calculated equilibrium state. In order to save computational time, the modified Newton-Raphson scheme is the method of first choice as long the structure behaves rather linearly. If the non-linearity increases, it may be advantageous to switch to the full Newton-Raphson scheme because of its faster convergence when compared to the modified Newton-Raphson scheme.

5.3 Arc-Length Control Technique

In the various Newton-Raphson schemes available, usually the load or the displacements are parameterized, i.e. increased incrementally. However, when the solution is close to limit points, convergence difficulties can occur. These difficulties can be circumvented if some path length parameter is used as independent parameter. These methods have been introduced first by Riks, see e.g. [107]. Generally one can rewrite (5.2) for a parameterized loading according to

$$ {}^{t+\Delta t}\lambda\boldsymbol{f} - {}^{t+\Delta t}\boldsymbol{r} = \boldsymbol{0} , \tag{5.14} $$

where ${}^{t+\Delta t}\lambda$ is the unknown scalar load multiplier. This parameter can increase or decrease depending on the state of the structure. For example, (5.13) can be rewritten as

$$ {}^{\tau}\boldsymbol{K}\Delta \boldsymbol{u}^{(s)} = ({}^{t}\lambda + \Delta\lambda^{(i)})\boldsymbol{f} - {}^{t+\Delta t}\boldsymbol{r}^{(s-1)} \ . \qquad (5.15) $$

Because now $n+1$ unknowns have to be determined, an additional constraint equation between the load multiplier λ and the displacements $\boldsymbol{u}$ is required, most generally given in the form

$$ f(\Delta\lambda^{(i)}, \Delta\boldsymbol{u}^{(i)}) = 0 \ . \qquad (5.16) $$

5.4 Linearized Buckling Analysis

The so-called fundamental path $0AC$ of a perfect shell structure is schematized in Fig. 5.1, its maximum load level is denoted as $\lambda_{\text{nl,p}}^{\text{lim}}$. Along this primary

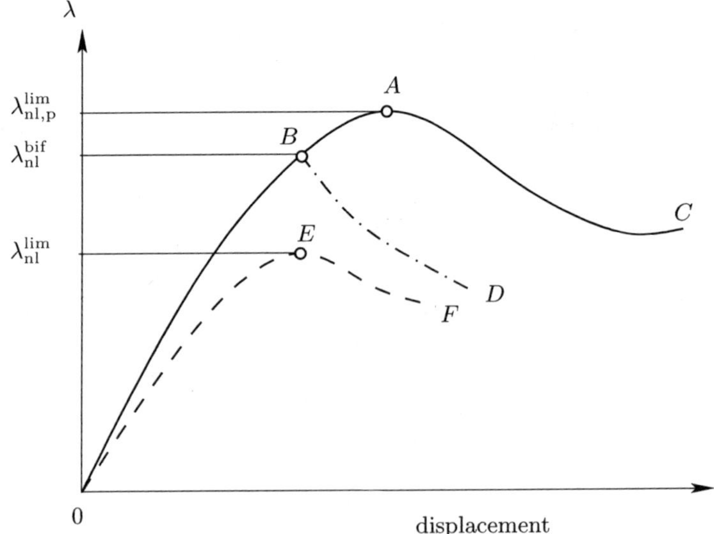

Fig. 5.1. Simplistic load-deflection curve showing limit and bifurcation points. Solid line: perfect shell structure (fundamental path); dashed line: imperfect shell shell structure; dashed-dotted line: post-buckling path [25]

path, there might be a bifurcation point at B, denoted as $\lambda_{\text{nl}}^{\text{bif}}$. If the fundamental path corresponds to axi-symmetric deformation and the post-buckling path BD corresponds to asymmetric deformation, than $\lambda_{\text{nl,p}}^{\text{lim}}$ of the perfect structure is of less engineering significance than $\lambda_{\text{nl}}^{\text{bif}}$. For real shell structures

having unavoidable imperfections, pure bifurcation buckling does not exist. In this case, the shell structure will follow the path $0EF$, a failure mode which is frequently denoted as snap-through at load level $\lambda_{\mathrm{nl}}^{\mathrm{lim}}$ in point E [25].
Limit points $\lambda_{\mathrm{nl}}^{\mathrm{lim}}$, see Fig. 5.1, can be computed by a general non-linear analysis using the arc-length control technique very accurately, depending on the tolerances of the iteration process. Also bifurcation points $\lambda_{\mathrm{nl}}^{\mathrm{bif}}$ can be detected because the stiffness matrix in this case is singular. However, the exact value of a bifurcation point is rather difficult to estimate by path following methods, because the only information one has is that between two subsequent states the sign of the determinant of the tangent stiffness matrix has changed. An accurate estimate of a bifurcation point can be obtained by carrying out a linearized buckling analysis based on a non-linear stress state of the structure. An additional advantage of such a buckling analysis is that it gives information of secondary solution paths before the bifurcation point has been passed. In a linearized buckling analysis, it is assumed that the tangent stiffness matrix between the states t and $t + \Delta t$ varies linearly according to

$$^{\tau}\boldsymbol{K} = {}^{t-\Delta t}\boldsymbol{K} + \lambda({}^{t}\boldsymbol{K} - {}^{t-\Delta t}\boldsymbol{K}) \ , \tag{5.17}$$

see e.g. [14]. A similar relation for the loading is assumed by

$$^{\tau}\boldsymbol{f} = {}^{t-\Delta t}\boldsymbol{f} + \lambda({}^{t}\boldsymbol{f} - {}^{t-\Delta t}\boldsymbol{f}) \ . \tag{5.18}$$

Because the tangent stiffness matrix at collapse or buckling is singular, its determinant must vanish

$$\det {}^{\tau}\boldsymbol{K} = 0 \ . \tag{5.19}$$

This condition is equivalent to the standard eigenvalue problem of the stiffness matrix where the right hand side is zero, i.e.

$$^{\tau}\boldsymbol{K}\boldsymbol{\phi}_i = \boldsymbol{0} \ , \tag{5.20}$$

where $\boldsymbol{\phi}$ is a non-zero eigenvector. Substituting (5.17) in (5.20) the standard eigenvalue problem is transformed into a generalized eigenvalue problem

$$^{t-\Delta t}\boldsymbol{K}\boldsymbol{\phi}_i = \lambda_i({}^{t-\Delta t}\boldsymbol{K} - {}^{t}\boldsymbol{K})\boldsymbol{\phi}_i \ , \tag{5.21}$$

where the matrix ${}^{t-\Delta t}\boldsymbol{K} - {}^{t}\boldsymbol{K}$ is in general indefinite (${}^{t-\Delta t}\boldsymbol{K}$ and ${}^{t}\boldsymbol{K}$ are still positive definite). Introducing

$$\gamma_i = \frac{\lambda_i - 1}{\lambda_i} \ , \tag{5.22}$$

(5.21) can be further simplified to

$$^{t}\boldsymbol{K}\boldsymbol{\phi} = \gamma_i \, {}^{t-\Delta t}\boldsymbol{K}\boldsymbol{\phi}, \tag{5.23}$$

where γ_i is always positive. Denoting the smallest eigenvalue as $\lambda_{\min}$ then the critical load (buckling or collapse) is given by

$$^{\tau}\boldsymbol{f}_{\mathrm{crit}} = {}^{t-\Delta t}\boldsymbol{f} + \lambda_{\min}({}^{t}\boldsymbol{f} - {}^{t-\Delta t}\boldsymbol{f}) \ . \tag{5.24}$$

6

Dynamic Analysis

6.1 System Equations

A general dynamical system subjected to an external force vector $\boldsymbol{f}(t)$ with a n-dimensional response $\boldsymbol{u}(t)$ satisfies the non-linear second-order differential equation

$$\boldsymbol{M}\ddot{\boldsymbol{u}} + \boldsymbol{r}(\boldsymbol{u}, \dot{\boldsymbol{u}}, \boldsymbol{q}) = \boldsymbol{f}(t) , \tag{6.1}$$

where "." derivatives with respect to time t and $\boldsymbol{r}(\boldsymbol{u}, \dot{\boldsymbol{u}}, \boldsymbol{q})$ is the non-linear restoring force, respectively. Where the time variability of the response quantities is obvious, it has not been explicitly stated in the subsequent derivations. The mass matrix $\boldsymbol{M}$ is assumed to be positive definite and time invariant. The internal restoring force vector $\boldsymbol{r}(\boldsymbol{u}, \dot{\boldsymbol{u}}, \boldsymbol{q})$ is a function of the displacements $\boldsymbol{u}$, the velocities $\dot{\boldsymbol{u}}$ and a set of variables $\boldsymbol{q}$, which describe the respective state of the non-linear elements. The evolution of these variables may be described by a first order differential equation, i.e.

$$\dot{\boldsymbol{q}} = \boldsymbol{g}(\boldsymbol{u}, \dot{\boldsymbol{u}}, \boldsymbol{q}) , \tag{6.2}$$

where the vector valued function $\boldsymbol{g}(.)$ is in general non-linear. Provided that the non-linearity of a system described by (6.1) does not exhibit instability-like behavior, the response $\boldsymbol{u}(t)$ can be computed – at least within certain time intervals – quite efficiently with a constant damping matrix $\boldsymbol{C}$ and a stiffness matrix $\boldsymbol{K}$, shifting the remaining non-linear restoring force vector to the right hand side of (6.1). For this purpose, it is assumed that the variables $\boldsymbol{q}$ are selected in a way, that $\boldsymbol{r}(\boldsymbol{u}, \dot{\boldsymbol{u}}, \boldsymbol{q})$ can be split into a linear force $\boldsymbol{C}\dot{\boldsymbol{u}} + \boldsymbol{K}\boldsymbol{u}$ and a non-linear force $\bar{\boldsymbol{r}}(\boldsymbol{u}, \dot{\boldsymbol{u}}, \boldsymbol{q})$, i.e.

$$\boldsymbol{r}(\boldsymbol{u}, \dot{\boldsymbol{u}}, \boldsymbol{q}) = \boldsymbol{C}\dot{\boldsymbol{u}} + \boldsymbol{K}\boldsymbol{u} + \bar{\boldsymbol{r}}(\boldsymbol{u}, \dot{\boldsymbol{u}}, \boldsymbol{q}) , \tag{6.3}$$

leading to

$$\boldsymbol{M}\ddot{\boldsymbol{u}} + \boldsymbol{C}\dot{\boldsymbol{u}} + \boldsymbol{K}\boldsymbol{u} = \boldsymbol{f}(t) - \bar{\boldsymbol{r}}(\boldsymbol{u}, \dot{\boldsymbol{u}}, \boldsymbol{q}) = \bar{\boldsymbol{f}}(t) . \tag{6.4}$$

It follows that $\bar{\boldsymbol{f}}(t)$ is a function of $(\boldsymbol{u}, \dot{\boldsymbol{u}}, \boldsymbol{q})$ so that iterative procedures are required in order to satisfy equilibrium conditions. It is obvious that the smaller the term $\bar{\boldsymbol{r}}(\boldsymbol{u}, \dot{\boldsymbol{u}}, \boldsymbol{q})$, the faster the convergence for satisfying the equilibrium conditions will be. Hence, the constant matrices $\boldsymbol{C}$ and $\boldsymbol{K}$ should be selected such that the vector $\bar{\boldsymbol{r}}(\boldsymbol{u}, \dot{\boldsymbol{u}}, \boldsymbol{q})$ remains small over a reasonable long time span. Equation (6.4) is especially well suited for cases where most of the elements of the structure remain linear and only a few reveal a non-linear behavior. In the following, it will be assumed that $\bar{\boldsymbol{r}}(\boldsymbol{u}, \dot{\boldsymbol{u}}, \boldsymbol{q})$ can be expressed by

$$\bar{\boldsymbol{r}}(\boldsymbol{u}, \dot{\boldsymbol{u}}, \boldsymbol{q}) = -\boldsymbol{R}\boldsymbol{q}(\boldsymbol{u}, \dot{\boldsymbol{u}}) \,. \tag{6.5}$$

Here, $\boldsymbol{R}$ denotes a constant matrix of size $n \times n_{\boldsymbol{q}}$, where $n_{\boldsymbol{q}}$ is the number of so-called auxiliary variables. For systems, where non-linearity can be neglected, (6.4) reduces to

$$\boldsymbol{M}\ddot{\boldsymbol{u}} + \boldsymbol{C}\dot{\boldsymbol{u}} + \boldsymbol{K}\boldsymbol{u} = \boldsymbol{f}(t) \,. \tag{6.6}$$

6.2 Time Integration

In a finite element analysis, (6.4) or (6.6) can be solved either in the original space by direct integration or in a reduced subspace by mode superposition. In the first approach, no transformation of the equation of motion is carried out prior to time integration. If the time span for which the response of the system has to be computed is relatively short, direct integration might be more efficient than mode superposition, since the time consuming solution of the general eigenvalue problem is not required. In addition, mode superposition methods approximate the response of the system due to mode truncation, which is not the case when direct integration is applied. In earthquake engineering, for example, the time span for which the response is calculated is in general in the range of 10-20 seconds, thus mode superposition tends to be in this field by far more efficient than direct integration.
For non-linear systems, unconditionally stable integration methods should be applied in order to ensure numerical stability. The Newmark integration scheme is a popular time integration method belonging to the class of implicit integration schemes. For specific values of integration parameters α and δ, see Sec. 6.2.1, it is unconditionally stable. Thus, the Newmark integration scheme is applied advantageously for solving (6.4) or (6.6) by both direct integration and the mode superposition (see Sec. 6.4), although any other appropriate time integration scheme may be used.

6.2.1 Newmark Method

Numerical integration in general implies a discretization of both the excitation and the response in time, ensuring static equilibrium for every time step. For the unconditionally stable Newmark method the integration parameters

$$\delta \geq 0.50 , \qquad \alpha \geq 0.25(0.5+\delta)^2 , \tag{6.7}$$

are usually taken as $\delta = 0.50$ and $\alpha = 0.25$, see e.g. [14]. The Newmark coefficients are defined by

$$a_0 = \frac{1}{\alpha \Delta t^2} , \qquad a_1 = \frac{\delta}{\alpha \Delta t} , \qquad a_2 = \frac{1}{\alpha \Delta t} , \qquad a_3 = \frac{1}{2\alpha} - 1 , \tag{6.8}$$

and

$$a_4 = \frac{\delta}{\alpha} - 1 , \qquad a_5 = \frac{\Delta t}{2}\left(\frac{\delta}{\alpha} - 2\right) , \qquad a_6 = \Delta t(1-\delta) , \qquad a_7 = \delta t . \tag{6.9}$$

The so-called effective stiffness matrix $\hat{\boldsymbol{K}}$ is given by

$$\hat{\boldsymbol{K}} = \boldsymbol{K} + a_0 \boldsymbol{M} + a_1 \boldsymbol{C} , \tag{6.10}$$

and is thus a function of the system matrices and the chosen parameters α, δ and Δt. For every time $t + \Delta t$ the effective load vector is calculated according to

$${}^{t+\Delta t}\hat{\boldsymbol{f}} = {}^{t+\Delta t}\boldsymbol{f} + \boldsymbol{M}(a_0\,{}^{t}\boldsymbol{u} + a_2\,{}^{t}\dot{\boldsymbol{u}} + a_3\,{}^{t}\ddot{\boldsymbol{u}}) + \boldsymbol{C}(a_1\,{}^{t}\boldsymbol{u} + a_4\,{}^{t}\dot{\boldsymbol{u}} + a_5\,{}^{t}\ddot{\boldsymbol{u}}) . \tag{6.11}$$

The displacements for time $t + \Delta t$ become available by solving

$$\hat{\boldsymbol{K}}\,{}^{t+\Delta t}\boldsymbol{u} = {}^{t+\Delta t}\hat{\boldsymbol{f}} . \tag{6.12}$$

As can be seen, (6.12) has to be solved for every time step, thus, as mentioned earlier, the Newmark integration scheme belongs to class of implicit integration schemes. When the displacements at time $t + \Delta t$ are known, the corresponding accelerations and the velocities can be determined by

$${}^{t+\Delta t}\ddot{\boldsymbol{u}} = a_0({}^{t+\Delta t}\boldsymbol{u} - {}^{t}\boldsymbol{u})) - a_2\,{}^{t}\dot{\boldsymbol{u}} - a_3\,{}^{t}\ddot{\boldsymbol{u}} , \tag{6.13}$$

and

$${}^{t+\Delta t}\dot{\boldsymbol{u}} = {}^{t}\boldsymbol{u} + a_6\,{}^{t}\ddot{\boldsymbol{u}} + a_7\,{}^{t+\Delta t}\ddot{\boldsymbol{u}} . \tag{6.14}$$

6.3 Mode Displacement Method

Mode superposition makes use of a transformation of the equation of motion onto a subspace prior to integration, thus being very efficient for the computation of the structural response of linear and – depending on the type of non-linearity – also for non-linear systems. Different model reduction techniques are available in the literature, see e.g. [77] for a more advanced projection technique. The equation of motion of a general dynamical system is given by, see e.g. [80, 20],

$$\boldsymbol{M}\ddot{\boldsymbol{u}}(t) + (\boldsymbol{G} + \boldsymbol{C})\dot{\boldsymbol{u}}(t) + (\boldsymbol{K} + \boldsymbol{H})\boldsymbol{u}(t) = \boldsymbol{f}(t) , \tag{6.15}$$

where $\boldsymbol{M}$, $\boldsymbol{C}$, $\boldsymbol{K}$ denote the symmetric mass, damping and stiffness matrices, respectively, i.e.

$$\boldsymbol{M} = \boldsymbol{M}^T, \qquad \boldsymbol{C} = \boldsymbol{C}^T, \qquad \boldsymbol{K} = \boldsymbol{K}^T , \tag{6.16}$$

and $\boldsymbol{G}$, $\boldsymbol{H}$ denote the skew symmetric gyroscopic and circulatory matrices, i.e.

$$\boldsymbol{G} = -\boldsymbol{G}^T, \qquad \boldsymbol{H} = -\boldsymbol{H}^T . \tag{6.17}$$

In case no gyroscopic and circulatory forces are present, (6.15) can be reduced to

$$\boldsymbol{M}\ddot{\boldsymbol{u}}(t) + \boldsymbol{C}\dot{\boldsymbol{u}}(t) + \boldsymbol{K}\boldsymbol{u}(t) = \boldsymbol{f}(t) . \tag{6.18}$$

Assuming a solution of (6.18) to have the form

$$\boldsymbol{u}(t) = \mathrm{e}^{st}\boldsymbol{\phi} , \tag{6.19}$$

where s is a constant scalar and $\boldsymbol{\phi}$ is a constant displacement vector, the generalized eigenvalue problem associated with (6.18) can be formulated as

$$\boldsymbol{K}\boldsymbol{\phi}_i = \lambda_i \boldsymbol{M}\boldsymbol{\phi}_i , \tag{6.20}$$

where

$$\lambda_i = -s_i^2 \qquad \text{or} \qquad s_i = \pm\sqrt{-\lambda_i} . \tag{6.21}$$

Vector $\boldsymbol{\phi}$ denotes the eigenvector associated with the scalar eigenvalue λ. The eigenvectors might be normalized such that the relations

$$\begin{aligned} \boldsymbol{\phi}_i^T \boldsymbol{M}\boldsymbol{\phi}_j &= \delta_{ij}, \qquad & \boldsymbol{\Phi}^T\boldsymbol{M}\boldsymbol{\Phi} &= \boldsymbol{I} , \\ \boldsymbol{\phi}_i^T \boldsymbol{K}\boldsymbol{\phi}_j &= \lambda_i\delta_{ij}, \qquad & \boldsymbol{\Phi}^T\boldsymbol{K}\boldsymbol{\Phi} &= \boldsymbol{\Lambda} , \end{aligned} \tag{6.22}$$

hold, with

$$\boldsymbol{\Phi} = [\boldsymbol{\phi}_1, \boldsymbol{\phi}_2, \ldots, \boldsymbol{\phi}_N] , \tag{6.23}$$

and

$$\boldsymbol{\Lambda} = \begin{bmatrix} \lambda_1 & 0 & \ldots & 0 \\ 0 & \lambda_2 & \ddots & \vdots \\ \vdots & \ddots & \ddots & 0 \\ 0 & \ldots & 0 & \lambda_N \end{bmatrix} , \tag{6.24}$$

where N is the number of modes used. In general, both the mass matrix $\boldsymbol{M}$ and the stiffness matrix $\boldsymbol{K}$ are positive definite, hence all eigenvalues are real and positive ($\lambda_i > 0$) and all eigenvectors $\boldsymbol{\phi}_i$ are real (see Chap. 3). In presence of rigid body modes, i.e. the stiffness matrix $\boldsymbol{K}$ is positive semi-definite, all eigenvalues are real and equal or larger than zero ($\lambda_i \geq 0$). Defining the eigenfrequencies of the system by $\omega_i = \sqrt{\lambda_i}$ it follows that

$$s_i = \pm j\omega_i \tag{6.25}$$

with $j = \sqrt{-1}$. For proportional damping (see e.g. [14]), i.e.

$$C = \alpha M + \beta K \ , \tag{6.26}$$

where α and β are constants, the damping matrix can be diagonalized by the eigenmodes according to

$$\phi_i C \phi_j = 2\omega_i \xi_i \delta_{ij} \ . \tag{6.27}$$

Here, ξ_i denotes the damping of the i-th mode. The equivalent modal damping for a given proportional damping can be determined by

$$\xi_i = \frac{\alpha + \beta\omega_i^2}{2\omega_i} \ . \tag{6.28}$$

Introducing the linear transformation

$$u = \sum_i z_i \phi_i \ , \tag{6.29}$$

where z denotes the vector of generalized or modal coordinates, the coupled system of equations given by (6.18) can be decoupled for proportional or modal damping. Using e.g. the Duhamel integral method, the general solution of (6.18) reads using (3.19), see e.g. [35],

$$\begin{aligned} z_i(t) =& \frac{1}{\omega_i'} \int_0^t \phi_i^T f(\tau) e^{-\xi_i \omega_i (t-\tau)} \sin(\omega_i'(t-\tau)) d\tau \\ &+ z_i(0) e^{-\xi_i \omega_i t} \cos(\omega_i' t) + \frac{1}{\omega_i'} \Big[\dot{z}_i(0) + \xi_i \omega_i z_i(0) \Big] e^{-\xi_i \omega_i t} \sin(\omega_i' t) \ , \end{aligned} \tag{6.30}$$

which simplifies for conservative systems to

$$z_i(t) = \frac{1}{\omega_i} \int_0^t \phi_i^T f(\tau) \sin(\omega_i(t-\tau)) d\tau + z_i(0) \cos(\omega_i t) + \frac{1}{\omega_i} \dot{z}_i(0) \sin(\omega_i' t) \ . \tag{6.31}$$

In (6.29), ω_i' is the circular eigenfrequency of the damped system given by $\omega_i' = \omega_i \sqrt{1 - \xi_i^2}$. The mode displacement method given by (6.29)–(6.30) or (6.31) may converge for large finite element systems rather slowly, i.e. a large number of modes has to be considered in (6.29). This is particularly true when stresses have to be evaluated, since in this case also higher modes contribute significantly to the overall response. The so-called mode acceleration method circumvents some of the drawbacks of the mode displacement method and will be reviewed in Sec. 6.4.

6.4 Mode Acceleration Method

The mode acceleration method again requires solutions of the general eigenvalue problem, see also (6.20). Using all n eigenvalues and eigenvectors, respectively, where n denotes the number of degrees of freedom of the structural system, it is possible to construct the flexibility matrix

$$\boldsymbol{K}^{-1} = \sum_{i=1}^{n} \frac{1}{\lambda_i} \boldsymbol{\phi}_i \boldsymbol{\phi}_i^T . \tag{6.32}$$

However, the advantage of mode superposition is the reduced set of $N << n$ modal coordinates. That is, generally, only a relatively small number N of modes will be considered in the analysis. In the mode displacement method, the contribution of the remaining modes to the structural response is ignored. In the mode acceleration method, however, their contribution to the overall response is approximated by their static response. This approach is justified because higher order modes react essentially in a static manner when excited by low frequencies.
Hence, the response consists of a dynamic part $\boldsymbol{u}_d(t)$ and a static part $\boldsymbol{u}_s(t)$, i.e.

$$\boldsymbol{u}(t) = \boldsymbol{u}_d(t) + \boldsymbol{u}_s(t) . \tag{6.33}$$

The dynamic modal response is represented by a linear combination of modal coordinates $\boldsymbol{z}(t)$,

$$\boldsymbol{u}_d(t) = \sum_{i=1}^{N} \boldsymbol{\phi}_i z_i(t) = \boldsymbol{\Phi}\boldsymbol{z}(t) , \tag{6.34}$$

and the remaining static part by the solution, see e.g. [33],

$$\boldsymbol{u}_s(t) = [K^{-1} - \sum_{i=1}^{N} \frac{1}{\lambda_i} \boldsymbol{\phi}_i \boldsymbol{\phi}_i^T] \bar{\boldsymbol{f}}(t) = [K^{-1} - \boldsymbol{\Phi}\boldsymbol{\Lambda}^{-1}\boldsymbol{\Phi}^T] \bar{\boldsymbol{f}}(t) . \tag{6.35}$$

Here,

$$\bar{\boldsymbol{f}}(t) = \boldsymbol{f}(t) + \boldsymbol{R}\boldsymbol{q}(t) , \tag{6.36}$$

denotes the loading as introduced in (6.4). The modal coordinates $\boldsymbol{z}(t)$ are determined by the solution of the decoupled single degree of freedom systems (e.g. Duhamel, Runge-Kutta, Newmark, etc.) using suitable modal damping ratios ζ, i.e.

$$\ddot{z}_i(t) + 2\zeta_i\omega_i\dot{z}_i(t) + \omega_i^2 z_i(t) = \boldsymbol{\phi}_i^T \bar{\boldsymbol{f}}(t) . \tag{6.37}$$

The static part of the response can be calculated by

$$\boldsymbol{u}_s(t) = \hat{\boldsymbol{B}}\boldsymbol{a}(t) + \hat{\boldsymbol{R}}\boldsymbol{q}(t) , \tag{6.38}$$

where the constant matrices $\hat{\boldsymbol{B}}$ and $\hat{\boldsymbol{R}}$ are specified by

$$\hat{\boldsymbol{B}} = \boldsymbol{K}^{-1}\boldsymbol{B} - \boldsymbol{\Phi}\boldsymbol{\Lambda}^{-1}\boldsymbol{\Phi}^T\boldsymbol{B}\,, \qquad \hat{\boldsymbol{R}} = \boldsymbol{K}^{-1}\boldsymbol{R} - \boldsymbol{\Phi}\boldsymbol{\Lambda}^{-1}\boldsymbol{\Phi}^T\boldsymbol{R}\,, \tag{6.39}$$

respectively and $\boldsymbol{B} = -\boldsymbol{M}\boldsymbol{I}_{\boldsymbol{a}}$. Equation (6.35) allows one to derive (6.39) which has to be computed only once. The fully populated matrix $\boldsymbol{K}^{-1}$ in (6.35) is not required in order to compute the matrices in (6.39). These matrices are computed by pre-multiplying (6.39) with $\boldsymbol{K}$ and solving the resulting equation by simple Gaussian elimination. The relations given above allow now to specify uniquely the state of the structural response by

$$\boldsymbol{u}(t) = \boldsymbol{\Phi}\boldsymbol{z}(t) + \hat{\boldsymbol{B}}\boldsymbol{a}(t) + \hat{\boldsymbol{R}}\boldsymbol{q}(t)\,, \tag{6.40}$$

and

$$\dot{\boldsymbol{u}}(t) = \frac{d}{dt}\boldsymbol{u}(t) = \boldsymbol{\Phi}\dot{\boldsymbol{z}}(t) + \hat{\boldsymbol{B}}\dot{\boldsymbol{a}}(t) + \hat{\boldsymbol{R}}\dot{\boldsymbol{q}}(t)\,. \tag{6.41}$$

Integration of the equations of motion for longer time spans in the subspace spanned by (6.40) and (6.41) is by far more efficient than direct integration of (6.4), where, for large finite element systems, more than three orders of magnitude of unknowns might by involved.

Part II

Probabilistic Methods and Procedures

7

Rational Treatment of Uncertainties

Structural analysis up to this date, generally is still based on a deterministic conception. Observed variations in loading conditions, material properties, geometry, etc. are taken into account by either selecting extremely high or low or average values, respectively, for representing the parameters. Hence, by this, uncertainties inherent in almost every analysis process are considered just intuitively. Observations and measurements of physical processes, however, clearly show their random characteristics. Statistical and probabilistic procedures provide a sound frame work for a rational treatment of analysis of these uncertainties. Moreover there are various types of uncertainties to be

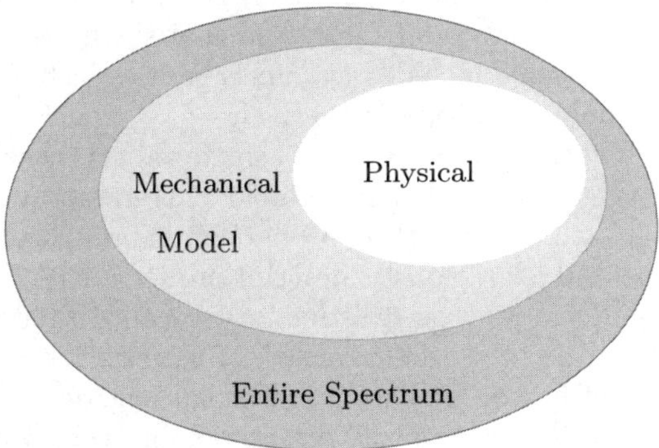

Fig. 7.1. Spectrum of Uncertainties

dealt with (see Fig. 7.1). While the uncertainties in mechanical modeling can be reduced as additional knowledge becomes available, the physical or intrinsic uncertainties can not. Furthermore, the entire spectrum of uncertainties is also not known. In reality, neither the true model nor the model parameters are deterministically known. Assuming that by finite element procedures structures and continua can be represented reasonably well, the question of the effect of the discretization still remains. It is generally expected, that an increase in the size of the structural models – in terms of degrees of freedom – will increase the level of realism of the model. Comparisons with measurements, however, clearly show that this expectation can not be confirmed. An ever refined finite element model just decreases the discretization error, but all other aspects contributing to the discrepancy between prediction and measurement will not be improved. There are several reasons for this. Among them is the fact that the finite element model, which is a mathematical idealization, represents the physical behavior not exactly but with a certain accuracy only. Typical examples for this are strongly non-linear interactions in a linear model, ignoring flexibilities at joints, inaccurate modeling of the boundary conditions, ignoring the non-linear interaction, etc. Furthermore, even if it is assumed that the idealized mathematical finite element model represents the structural behavior, the model parameters do show uncertainties. As already stated above, these uncertainties refer to both loading – environmental loading such as water waves, wind, earthquakes, etc. are good examples for this – as well as to structural properties, such as imperfections of geometry, thickness, Young's modulus, material strength, fracture toughness, damping characteristics, etc. It is also well known, that the results of experimental measurements are subjected to uncontrollable random effects. This is the main reason, why they are so difficult to reproduce. This fact leads directly to the claim as made before, i.e. that an increase in the number of degrees of freedom does not compensate for the insufficient modeling of physical phenomena, such as not taking into account the uncertainties in the boundary conditions, etc. Needless to say that it is most important that the model reflects physical phenomena. This of course includes the uncertainties in both the structural properties and loading conditions respectively.

While in the deterministic conception a single value is considered to suffice for the representation of a particular variable, it is in fact a great number of values – each associated with a certain probability of occurrence of a particular value – which is needed for a realistic description, see Fig. 7.2. Hence the variables in their basic form may be described as so-called random variables X. Typical examples are e.g. the yield strength of materials. etc. The associated uncertainty are quantified by probability measures, e.g. described as probability density functions. In other words, the probability that a parameter takes on values within an interval is,

$$P(a < X \leq b) = \int_a^b f(x)\, dx \; . \tag{7.1}$$

This one dimensional definition certainly can be expanded easily for multi dimensional cases. The distribution of the occurrence of the various values i.e. $f(x)$, also denoted as the probability density function, is generally characterized by certain types of function such as the normal or Gaussian distribution, etc. The parameters, such as the central tendency or mean value as well as the variance, are estimated by statistical procedures. For time variant or stationary processes, the probability density refers not only to one time instant, but to other times as well i.e. to a family of random variables $X(t_1)$, $X(t_2)$ more simply denoted by $X(t)$. Again, if the distribution of $X(t_1)$, $X(t_2)$ are of Gaussian characteristics, such a process is denoted as Gaussian stochastic process, see Sec. 7.1. Typical examples are wind, wave, earthquake records, etc.

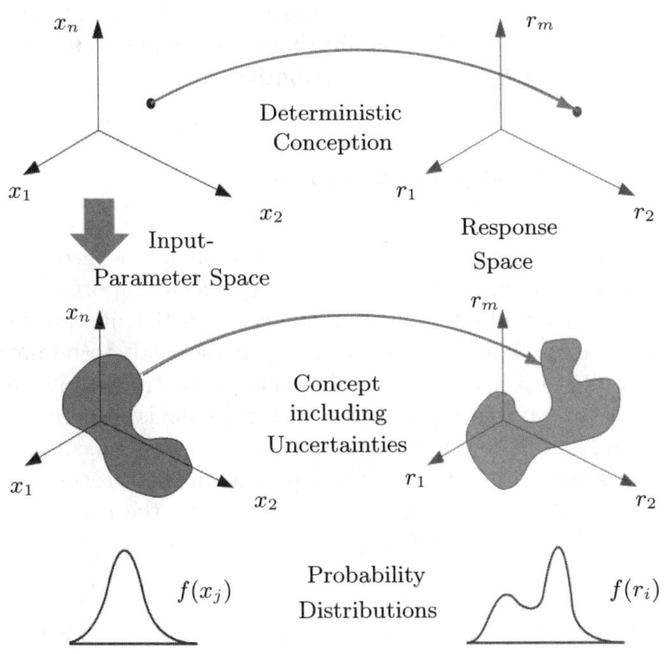

Fig. 7.2. Deterministic Conception versus Concept including Uncertainties

Finite element models generally contain quite a large number of parameters like elasticity constants, geometry specifications, loading parameters, boundary conditions, etc., of which most values are not perfectly known. It was already stated above, that the so called "true" parameters can, if at all, be determined in exceptional cases only, i.e. by experiments. Hence the values used in deterministic finite element analysis are so called nominal values which deviate to a certain extent from the unknown true value. The uncertainties within the input parameters naturally result in uncertainties of the output, i.e. the response. Since response predictions are the central goal of any finite element analysis, and all predictions depend more or less on the uncertain input parameters, a rational approach has to include these unavoidable uncertainties.

Data are always scarce. Accessible statistical information, if any, might be restricted to the mean value, the standard deviation, upper and lower fractile values or upper and lower bounds. However, with this concept, whatever information is available, it can be used and – as new information or data becomes available – updated. Under these circumstances, it is reasonable to select the most convenient distribution which reflects the known or assumed variability or uncertainty and avoids realizations which are not physically meaningful. For many reasons, the Gaussian normal distribution and sometimes the log-normal distribution respectively are preferred. Since the uncertain input is specified mathematically by probability laws, the response follows also such laws, i.e. has a well defined unique distribution.

7.1 Gaussian Stochastic Process

The Gaussian stochastic process plays a central role in the theory as well as in the application of probabilistic methods. Its significant importance is based on the well known Central Limit theorem, which states that if a physical process is the result of many additive effects, that process tends to be normally distributed. But also in cases, where the Gaussian process represents only a rough approximation for a physical phenomenon, this model is the starting point for many non-Gaussian models of stochastic processes. A stochastic process $x(t)$, see Chap. 8, is called a Gaussian or normal stochastic process, if, for every finite set $\{t_1, t_2, ...t_n\}$ in $D \subseteq \mathcal{R}^1$, where $D = [0, T]$, the n random variables $x(t_1), x(t_2), ..., x(t_n)$ have the joint probability density function according to

$$\begin{gathered} f_n(x_1, ...x_n; t_1, ..., t_2) = \\ (2\pi)^{(-n/2)} |\boldsymbol{\Gamma}_{\boldsymbol{xx}}(t)|^{-1/2} \mathrm{e}^{-\frac{1}{2}(\boldsymbol{x}-\boldsymbol{\mu}_{\boldsymbol{x}}(t))^T \boldsymbol{\Gamma}_{\boldsymbol{xx}}(t)^{-1}(\boldsymbol{x}-\boldsymbol{\mu}_{\boldsymbol{x}}(t))} , \end{gathered} \tag{7.2}$$

where $\boldsymbol{x}^T = [x_1, x_2, ...x_n]$, $\boldsymbol{\mu}_{\boldsymbol{x}}(t)$ is the mean vector

$$\boldsymbol{\mu}_{\boldsymbol{x}}(t) = E\{\boldsymbol{x}(t)\} \tag{7.3}$$

and $\boldsymbol{\Gamma}_{\boldsymbol{xx}}(t)$ is the covariance matrix, whose elements Γ_{ij} are given by

$$\Gamma_{ij}(t) = E\{(x(t_i) - \mu_{x_i}(t_i))(x(t_j) - \mu_{x_i}(t_j))\} . \tag{7.4}$$

The expectation operator $E\{\cdot\}$ of a continuous process is defined by

$$E\{g(x(t))\} = \int_{-\infty}^{\infty} g(x) f_g(x, t) \,\mathrm{d}x , \tag{7.5}$$

if the improper integral is absolutely convergent, i.e.

$$\int_{-\infty}^{\infty} g(x) f_g(x, t) \,\mathrm{d}x < \infty . \tag{7.6}$$

Also from the theoretical point of view, the Gaussian process has some very important properties, see e.g. [141]:

(i) The marginal or first order probability density function of a stochastic process $x(t)$ at a fixed time t has the form

$$f_1(x,t) = \frac{\partial F_1(x,t)}{\partial x} = (\frac{1}{\sqrt{2\pi}\sigma_x} \mathrm{e}^{\left[(x-\mu_x)^2/(2\sigma_x^2)\right]} . \tag{7.7}$$

The corresponding first order cumulative distribution function $F_1(x,t)$ is given by

$$F_1(x,t) = \frac{1}{\sqrt{2\pi}\sigma_x} \int_{\infty}^{x} \mathrm{e}^{\left[(u-\mu_x)^2/(2\sigma_x^2)\right]} \, \mathrm{d}u . \tag{7.8}$$

(ii) A Gaussian stochastic process is completely specified by its mean function $\mu_x(t)$ and covariance function $\gamma_{xx}(t)$. The n-th moment of a general stochastic process is defined by

$$\mu_x^n(t) = E\{x^n(t)\} = \int_{-\infty}^{\infty} x^n f_1(x,t) \, \mathrm{d}x . \tag{7.9}$$

The n-th central moment is given by

$$\gamma_x^n(t) = E\{(x(t) - \mu_x(t))^n\} = \int_{-\infty}^{\infty} (x - \mu_x(t))^n f_1(x,t) \, \mathrm{d}x , \tag{7.10}$$

where $\mu_x^1(t) := \mu_x(t)$. For a Gaussian stochastic process the central moments are

$$\mu_x^n = \begin{cases} 0 & \text{if } n \text{ is odd} , \\ 1 \cdot 3 \cdots (n-1)\sigma_x^n & \text{if } n \text{ is even} . \end{cases} \tag{7.11}$$

The nm-th joint central moment is given by

$$\begin{aligned} \gamma_{xx}^{nm}(t_1,t_2) &= E\{(x(t_1) - \mu_x(t_1))^n (x(t_2) - \mu_x(t_2))^m\} = \\ &\int_{-\infty}^{\infty} (x_1 - \mu_x(t_1))^n (x_2 - \mu_x(t_2))^m f_2(x_1,x_2;t_1,t_2) \, \mathrm{d}x . \end{aligned} \tag{7.12}$$

As a consequence for Gaussian stochastic processes, all odd-order joint central moments vanish.

(iii) Weak stationarity implies strict stationarity.

(iv) A Gaussian stochastic process remains Gaussian under linear transformations, a property that plays a central role in the procedure described in Chap. 11.

7.2 Representation of Stochastic Processes

Many physical quantities involving random fluctuations in time and space might be adequately described by stochastic processes, fields and waves. Typical examples of engineering interest are earthquake ground motion, sea waves,

wind turbulence, road roughness, imperfection of shells, fluctuating properties in random media, etc. For these quantities, probabilistic characteristics of the process are known from various measurements and investigations in the past. In structural engineering, the available probabilistic characteristics of random quantities affecting the loading or the mechanical system can be often not utilized directly to account for the randomness of the structural response due to its complexity. For example, in the case of stability analyses of thin shell structures, the structural response will be in general non-linear exhibiting critical points along the solution path and it might be too difficult to compute the probabilistic characteristics of the response by other means than Monte Carlo simulation. For the purpose of Monte Carlo simulation sample functions of the involved stochastic process must be generated. These sample functions should represent accurately the characteristics of the underlying stochastic process or fields and might be stationary and non-stationary, homogeneous or non-homogeneous, one-dimensional or multi-dimensional, uni-variate or multivariate, Gaussian or non-Gaussian, depending very much on the requirements of accuracy of realistic representation of the physical behavior and on the available statistical data.
The main requirement on the sample function is its accurate representation of the available stochastic information of the process. The associated mathematical model can be selected in any convenient manner as long it reproduces the required stochastic properties. Therfore, quite different representations have been developed and might be utilized for this purpose. The most common representations for stochastic processes are:

(i) Spectral representation,
(ii) Karhunen-Loève representation and polynomial chaos representation,
(iii) Wavelets representation.

7.2.1 Spectral Representation

Among the various methods listed above, the spectral representation appears to be most widely used (see e.g. [105, 134, 128]). According to this procedure, samples with specified power spectral density information are generated. For the stationary or homogeneous case the fast Fourier transform techniques is utilized for a dramatic improvement of its computational efficiency (see e.g. [161, 162]). Recent advances in this field provide efficient procedures for the generation of two-dimensional and three-dimensional homogeneous Gaussian stochastic fields using the fast Fourier transform technique (see e.g. [129, 130, 131, 132, 133]). The spectral representation method generates ergodic sample functions of which each fulfills exactly the requirements of a target power spectrum. These procedures can be extended to the non-stationary case, to the generation of stochastic waves (see e.g. [40]) and to incorporate non-Gaussian stochastic fields (see e.g. [160]) by a memoryless nonlinear transformation together with an iterative procedure to meet the target spectral density.

7.2.2 Karhunen-Loève Representation and Polynomial Chaos Representation

A quite general representation utilized for Gaussian stochastic processes and fields is the Karhunen-Loève expansion of the covariance function, see Chap. 8 for a thorough treatment of this subject. A generalization of the Karhunen-Loève expansion denoted as polynomial chaos expansion has been proposed for applications where the covariance function is not known a priori (see [27, 50, 47, 111, 44]). These polynomials are orthogonal. For the special case of a Gaussian random process the polynomial chaos representation coincides with the Karhunen-Loève expansion. The polynomial chaos expansion is adjustable in two ways: Increasing the number of random coefficients results in a refinement of the random fluctuations, while an increase of the maximum order of the polynomial captures non-linear (non-Gaussian) behavior of the process. However, the problem when dealing with the polynomial chaos expansion is the large number of terms required to represent strong non-Gaussian processes.

7.2.3 Wavelets Representation

The spectral representation by Fourier analysis is not well suited to describe local feature in the time or space domain. This disadvantage is overcome in wavelets analysis which provides an alternative of breaking a signal down into its constituent parts. This representation has found important applications in signal processing (see e.g. [38, 85]) and it has been recently utilized in stochastic mechanics for representing e.g. random sea motion [154] and random fields [165]. For a large class of stochastic problems, wavelet expansion can be considered as an approximate Karhunen-Loève expansion. For system identification and damage detection purposes, wavelet analysis may prove to be an advantageous approach.

7.2.4 Non-Gaussian Models

In some cases of applications the physics or data might be inconsistent with the Gaussian distribution. For such cases, non-Gaussian models have been developed employing various concepts to meet the desired target distribution as well as the target correlation structure (spectral density). Certainly the most straight forward procedures is the memoryless non-linear transformation of Gaussian processes (e.g. [160]) utilizing the spectral representation. An alternative approach utilizes non-linear filters to represent non-Gaussian processes and fields excited by Gaussian white noise. Non-linear filters are utilized in the recent works to generate a stationary non-Gaussian stochastic process in agreement with a given first-order probability density function and the spectral density [26]. For generating samples of the non-linear filter represented by a stochastic differential equations, well developed numerical procedures are available (see e.g. [68]). For an extensive overview of non-Gaussian data, mathematical models generating non-Gaussian processes, and classes of non-Gaussian processes it is referred to e.g. [55].

8

Karhunen-Loève Expansion

In the following, second order stochastic processes are considered, i.e. random functions $x(t)$ satisfying

$$E\{x(t)^2\} < \infty , \qquad t \in D \subseteq \mathcal{R}^1 , \tag{8.1}$$

where t is a temporal or spatial parameter, and $D = [0, T]$ denotes the interval over which $x(t)$ is defined. With regard to uncertainty analysis, the most important characteristics of a stochastic process $x(t)$ are captured by its mean function

$$\mu_x(t) = E\{x(t)\} \tag{8.2}$$

and its covariance function

$$\Gamma_{xx}(t, s) = E\{(x(t) - \mu_x(t))(x(s) - \mu_x(s))\} . \tag{8.3}$$

Among the various definitions of stochastic processes available, the parametric one can be seen as the basis for the group of orthogonal series representations of a stochastic process. Following this definition, a stochastic process can be described by an explicit analytical expression involving an at-most countable set of random variables as parameters

$$x(t) = \phi(t; \zeta_1, \zeta_2, ...) , \tag{8.4}$$

in which ϕ is a specified deterministic function of t and $\{\zeta_1, \zeta_2, ...\}$ denotes a countable set of random variables. The probability law of $x(t)$ is fully defined by the joint probability distribution of these random variables together with the functional form of $\phi(t)$ [141]. Being more precise, a stochastic process can be represented in terms of orthonormal functions $\psi(t)$ according to

$$x(t) = \sum_{i=1}^{\infty} \alpha_i \psi_i(t) , \tag{8.5}$$

where the $\psi(t)$'s are satisfying the relation

$$\int_D \psi_i(t)\psi_j(t)\,\mathrm{d}t = \delta_{ij}\,. \tag{8.6}$$

The random coefficients α can be calculated according to the Fourier series theorem

$$\alpha_i = \int_D x(t)\psi_i(t)\,\mathrm{d}t\,. \tag{8.7}$$

The mean values μ_α and the covariances $\sigma_{\alpha\alpha}$ of the random coefficients are given by

$$\mu_{\alpha_i} = \int_D \mu_x(t)\psi_i(t)\,\mathrm{d}t \tag{8.8}$$

and

$$\sigma_{\alpha_i\alpha_j} = \int_D\int_D \Gamma_{xx}(t,s)\psi_i(t)\psi_j(s)\,\mathrm{d}t\,\mathrm{d}s\,. \tag{8.9}$$

For zero mean Gaussian stochastic processes, the random coefficients are Gaussian random variables with mean values equal to zero, which are, depending on the orthonormal functions used, not necessarily uncorrelated [142]. Several orthogonal series expansions of stochastic processes are documented in the literature, e.g. the random trigonometric polynomials [41, 54], which belong to the class orthogonal series expansions with correlated random coefficients. In recent years the so-called Karhunen-Loève expansion, see e.g. [39, 73, 50, 147, 11, 93, 92], has quite often been used in stochastic mechanics because of its attractive features:

(i) The mean-square error resulting from a finite representation of a stochastic the process is minimized. In other words, the eigenfunctions of the Karhunen-Loève expansion are adapted in a way that they allow a most efficient representation of this covariance kernel.
(ii) The Karhunen-Loève expansion represents Gaussian stationary as well as non-stationary processes, respectively. This is also true – but not generally – for a finite number of terms in (8.5).
(iii) For Gaussian stochastic processes the random coefficients α are uncorrelated and thus independent random variables. This property makes the Karhunen-Loève expansion very useful for generating samples of Gaussian stochastic processes.

For stochastic processes of arbitrary marginal probability density functions, the Karhunen-Loève expansion is still valid, although in this case the random variables are not uncorrelated anymore [48].

The stochastic process $x(t)$ has been considered in the preceding discussion to be continuous, i.e. the domain over which $x(t)$ is defined corresponds to infinite dimensional Hilbert spaces. However, due to a necessary discretization of the parameter t in computational mechanics, only discrete stochastic processes are treated in this monograph, implying that the discrete

version of the Karhunen-Loève expansion – in the following denoted as discrete Karhunen-Loève expansion – is applied. Quite frequently, the discrete Karhunen-Loève expansion is also called principal component analysis. In the subsequent sections, the discrete Karhunen-Loève expansion is discussed for scalar stochastic processes, vector stochastic processes and random fields.
In order to keep notations clear, the continuous Karhunen-Loève expansion is exclusively used in the derivations for the described procedures in the subsequent chapters.

8.1 Scalar Stochastic Processes

A continuous stochastic process $x(t)$ can be discretized (sampled) at ordinarily equal intervals Δt, yielding a vector $\boldsymbol{x}$ whose elements are related to the value of $x(t)$ at a certain t, i.e.

$$\boldsymbol{x} = \begin{Bmatrix} x(0) \\ x(1\Delta t) \\ x(2\Delta t) \\ \vdots \\ x(T) \end{Bmatrix} . \tag{8.10}$$

Using the relation $t_k = k\Delta t$, the following notation will be used

$$x(t_k) = \boldsymbol{x}[k] \,, \qquad x(0 \le t \le T) = \boldsymbol{x}[:] = \boldsymbol{x} \,, \tag{8.11}$$

where the "colon" specifies all elements of a vector or a column or row of a matrix, see e.g. [51]. The Karhunen-Loève expansion of a scalar discrete Gaussian second order stochastic process $\boldsymbol{x}$, is defined as

$$\boldsymbol{x} = \boldsymbol{\mu}_{\boldsymbol{x}} + \sum_{j=1}^{N_{\boldsymbol{x}}} \xi_j \sqrt{\lambda_j} \boldsymbol{\psi}_j \equiv \boldsymbol{x}^{(0)} + \sum_{j=1}^{N_{\boldsymbol{x}}} \xi_j \boldsymbol{x}^{(j)} \,, \tag{8.12}$$

where $\boldsymbol{\mu}_x$ and $N_{\boldsymbol{x}} = T/\Delta t + 1$ denote the mean vector and the number of elements of $\boldsymbol{x}$, respectively. Let $\boldsymbol{x}_{(r)}, r = 1, 2, ..., n$ denote the r-th realization of the stochastic process $\boldsymbol{x}$. Then the covariance matrix $\boldsymbol{\Gamma}_{\boldsymbol{xx}}$ of $\boldsymbol{x}$ is defined according to

$$\boldsymbol{\Gamma}_{\boldsymbol{xx}} = E\{(\boldsymbol{x} - \mu_{\boldsymbol{x}})(\boldsymbol{x} - \mu_{\boldsymbol{x}})^T\} = \lim_{n\to\infty} \frac{1}{n} \sum_{r=1}^{n} (\boldsymbol{x}_{(r)} - \mu_{\boldsymbol{x}_{(r)}})(\boldsymbol{x}_{(r)} - \mu_{\boldsymbol{x}_{(r)}})^T\} \,. \tag{8.13}$$

The Karhunen-Loève vectors $\boldsymbol{x}^{(j)} = \sqrt{\lambda_j} \boldsymbol{\psi}_j$ associated with $\boldsymbol{x}$ are determined by solving the algebraic eigenvalue problem

$$\boldsymbol{\Gamma}_{\boldsymbol{xx}} \boldsymbol{\psi}_j = \lambda_j \boldsymbol{\psi}_j \,. \tag{8.14}$$

The orthonormality relation for eigenvectors $\boldsymbol{\psi}$ is given by

$$\boldsymbol{\psi}_i^T \boldsymbol{\psi}_j = \delta_{ij} \; . \tag{8.15}$$

The Gaussian random variables ξ as used in (8.12) are defined according to

$$\xi_j = \frac{1}{\sqrt{\lambda_j}} (\boldsymbol{x} - \boldsymbol{\mu}_{\boldsymbol{x}})^T \boldsymbol{\psi}_j \; , \tag{8.16}$$

a relation which holds because of

$$\xi_j = \frac{1}{\sqrt{\lambda_j}} \Big(\sum_{l=1}^{N_{\boldsymbol{x}}} \xi_l \sqrt{\lambda_l} \boldsymbol{\psi}_l \Big)^T \boldsymbol{\psi}_j \; . \tag{8.17}$$

In addition, (8.16) defines the random variables $\xi_{(r),i}$ corresponding to a specific realization of $\boldsymbol{x}$, i.e. $\boldsymbol{x}_{(r)}$,

$$\xi_{(r),j} = \frac{1}{\sqrt{\lambda_j}} (\boldsymbol{x}_{(r)} - \boldsymbol{\mu}_{\boldsymbol{x}_{(r)}})^T \boldsymbol{\psi}_j \; . \tag{8.18}$$

As previously mentioned, the Gaussian random variables ξ have the properties

$$E\{\xi_i\} = 0 \; , \qquad E\{\xi_i \xi_j\} = \delta_{ij} \; . \tag{8.19}$$

The first one follows from $E\{\boldsymbol{x}\} = \boldsymbol{\mu}_x$, while the second one makes use of the relation

$$\xi_j^T = \xi_j = \frac{1}{\sqrt{\lambda_j}} \boldsymbol{\psi}_j^T (\boldsymbol{x} - \boldsymbol{\mu}_{\boldsymbol{x}}) \; . \tag{8.20}$$

Substituting (8.16) and (8.20) in the second part of (8.19) one obtains

$$E\{\xi_j \xi_l\} = E\Big\{ \frac{1}{\sqrt{\lambda_j}} \boldsymbol{\psi}_j^T (\boldsymbol{x} - \boldsymbol{\mu}_{\boldsymbol{x}}) \frac{1}{\sqrt{\lambda_l}} (\boldsymbol{x} - \boldsymbol{\mu}_{\boldsymbol{x}})^T \boldsymbol{\psi}_l \Big\} \; , \tag{8.21}$$

which can be, using (8.13) and (8.14), simplified to

$$E\{\xi_j \xi_l\} = E\Big\{ \frac{1}{\sqrt{\lambda_j}} \frac{1}{\sqrt{\lambda_l}} \boldsymbol{\psi}_j^T \boldsymbol{\Gamma}_{\boldsymbol{xx}} \boldsymbol{\psi}_l \Big\} = E\Big\{ \frac{1}{\sqrt{\lambda_j}} \frac{1}{\sqrt{\lambda_l}} \boldsymbol{\psi}_j^T \lambda_l \boldsymbol{\psi}_l \Big\} = \delta_{jl} \; . \tag{8.22}$$

The mean $\boldsymbol{\mu}[k]$ of an element $\boldsymbol{x}[k]$ of $\boldsymbol{x}$ can be calculated by

$$\boldsymbol{\mu}[k] = \boldsymbol{x}^{(0)}[k] \; . \tag{8.23}$$

With regard to (8.13), the covariance of two elements $\boldsymbol{x}[k]$ and $\boldsymbol{x}[l]$ is given by

$$\boldsymbol{\Gamma}_{\boldsymbol{xx}}[k,l] = E\Big\{ \sum_{i=1}^{N_{\boldsymbol{x}}} \xi_i \boldsymbol{x}^{(i)}[k] \sum_{j=1}^{N_{\boldsymbol{x}}} \xi_j \boldsymbol{x}^{(j)}[l] \Big\} = \sum_{j=1}^{N_{\boldsymbol{x}}} \boldsymbol{x}^{(j)}[k] \boldsymbol{x}^{(j)}[l] \; , \tag{8.24}$$

in particular

$$\boldsymbol{\Gamma}_{\boldsymbol{xx}}[k,k] = \boldsymbol{\sigma}_{\boldsymbol{x}}^2[k] = \sum_{j=1}^{N_{\boldsymbol{x}}} \boldsymbol{x}[k]^{(j)^2} . \tag{8.25}$$

The Karhunen-Loève representation of the full covariance matrix is thus given by

$$\boldsymbol{\Gamma}_{\boldsymbol{xx}} = \sum_{j=1}^{N_{\boldsymbol{x}}} \boldsymbol{x}^{(j)} \boldsymbol{x}^{(j)T} . \tag{8.26}$$

Second moment characteristics of a linear combination

$$z = \sum_k c_k \boldsymbol{x}[k] \tag{8.27}$$

with constants c_k can be calculated by

$$\mu_z = \sum_k c_k \boldsymbol{x}^{(0)}[k] \tag{8.28}$$

and

$$\sigma_z^2 = \sum_{j=1}^{N_{\boldsymbol{x}}} \Big(\sum_k c_k \boldsymbol{x}^{(j)}[k] \Big)^2 . \tag{8.29}$$

The covariance of two linear combinations

$$z_1 = \sum_i c_i \boldsymbol{x}[i] \, , \qquad z_2 = \sum_k c_k \boldsymbol{x}[k] \tag{8.30}$$

with constants c_i and c_k, respectively, is given by

$$\sigma_{z_1 z_2} = \sum_{j=1}^{N_{\boldsymbol{x}}} \Big(\sum_i c_i \boldsymbol{x}^{(j)}[i] \Big) \Big(\sum_k c_k \boldsymbol{x}^{(j)}[k] \Big) . \tag{8.31}$$

8.2 Vector Stochastic Processes

So far, the theory of the discrete Karhunen-Loève expansion has been applied to a scalar stochastic process $\boldsymbol{x}$, after discretizing a continuous process $x(t)$ with respect to the parameter t. However, when dealing with the Karhunen-Loève expansion in context with finite element systems as described in Chap. 6, this theory has to be extended to vector stochastic processes. A m-dimensional continuous vector stochastic process $\boldsymbol{x}(t)$ with the components $x_1(t), x_2(t), ..., x_m(t)$ is defined by

$$\boldsymbol{x}(t) = \begin{Bmatrix} x_1(t) \\ x_2(t) \\ \vdots \\ x_m(t) \end{Bmatrix} . \tag{8.32}$$

For vector stochastic processes, the second order information can be related to the correlation between different components of $\boldsymbol{x}(t)$ at the same time or location t

$$\boldsymbol{\Gamma}_{\boldsymbol{xx}}(t) = E\{(\boldsymbol{x}(t) - \mu_{\boldsymbol{x}}(t))(\boldsymbol{x}(t) - \mu_{\boldsymbol{x}}(t))^T\}\ , \tag{8.33}$$

and the correlation between different components between different times or locations

$$\boldsymbol{\Gamma}_{\boldsymbol{xx}}(t,s) = E\{(\boldsymbol{x}(t) - \mu_{\boldsymbol{x}}(t))(\boldsymbol{x}(s) - \mu_{\boldsymbol{x}}(s))^T\}\ , \tag{8.34}$$

being both functions of $m \times m$ matrices. Similar and in order to preserve the previously defined notation, a discrete vector stochastic process is defined by

$$\boldsymbol{x}[:,k] = \begin{Bmatrix} \boldsymbol{x}_1[k] \\ \boldsymbol{x}_2[k] \\ \vdots \\ \boldsymbol{x}_m[k] \end{Bmatrix} = \begin{Bmatrix} x_1(k\Delta t) \\ x_2(k\Delta t) \\ \vdots \\ x_m(k\Delta t) \end{Bmatrix}\ , \tag{8.35}$$

where $\boldsymbol{x}$ represents a matrix of size $m \times N_{\boldsymbol{x}}$ ($N_{\boldsymbol{x}} = T/\Delta t + 1$, see above). The discrete counterparts of (8.33) and (8.34) are represented by three- and four-dimensional matrices of sizes $m \times m \times N_{\boldsymbol{x}}$ and $m \times m \times N_{\boldsymbol{x}} \times N_{\boldsymbol{x}}$, respectively, i.e.

$$\boldsymbol{\Gamma}_{\boldsymbol{xx}}[:,:,k] = E\{(\boldsymbol{x}[:,k] - \mu_{\boldsymbol{x}}[:,k])(\boldsymbol{x}[:,k] - \mu_{\boldsymbol{x}}[:,k])^T\} \tag{8.36}$$

and

$$\boldsymbol{\Gamma}_{\boldsymbol{xx}}[:,:,k,l] = E\{(\boldsymbol{x}[:,k] - \mu_{\boldsymbol{x}}[:,k])(\boldsymbol{x}[:l] - \mu_{\boldsymbol{x}}[:,l])^T\}\ . \tag{8.37}$$

The Karhunen-Loève representation of discrete vector stochastic processes can be obtained by reshaping the vector process $\boldsymbol{x}$ to a one dimensional (scalar) process $\boldsymbol{X}$ according to

$$\boldsymbol{X} = \begin{Bmatrix} \boldsymbol{x}_1 \\ \boldsymbol{x}_2 \\ \vdots \\ \boldsymbol{x}_m \end{Bmatrix}\ . \tag{8.38}$$

Then the covariance matrix $\boldsymbol{\Gamma}_{\boldsymbol{XX}}$ is defined by

$$\boldsymbol{\Gamma}_{\boldsymbol{XX}} = E\{(\boldsymbol{X} - \mu_{\boldsymbol{X}})(\boldsymbol{X} - \mu_{\boldsymbol{X}})^T\}\ , \tag{8.39}$$

having the structure

$$\boldsymbol{\Gamma}_{\boldsymbol{XX}} = \begin{bmatrix} \boldsymbol{\Gamma}_{\boldsymbol{x}_1\boldsymbol{x}_1} & \cdots & \boldsymbol{\Gamma}_{\boldsymbol{x}_1\boldsymbol{x}_m} \\ \vdots & \ddots & \vdots \\ \boldsymbol{\Gamma}_{\boldsymbol{x}_m\boldsymbol{x}_1} & \cdots & \boldsymbol{\Gamma}_{\boldsymbol{x}_m\boldsymbol{x}_m} \end{bmatrix}\ . \tag{8.40}$$

Matrix $\boldsymbol{\Gamma}_{\boldsymbol{XX}}$ is of size $N_{\boldsymbol{X}} \times N_{\boldsymbol{X}}$ where $N_{\boldsymbol{X}} = mN_{\boldsymbol{x}}$ and the corresponding Karhunen-Loève vectors are defined by

$$\boldsymbol{\Gamma}_{\boldsymbol{XX}}\boldsymbol{\psi}_j = \lambda_j \boldsymbol{\psi}_j \,. \tag{8.41}$$

The Karhunen-Loève representation of $\boldsymbol{X}$ then reads

$$\boldsymbol{X} = \boldsymbol{\mu}_{\boldsymbol{X}} + \sum_{j=1}^{N_{\boldsymbol{X}}} \xi_j \sqrt{\lambda_j}\boldsymbol{\psi}_j \equiv \boldsymbol{X}^{(0)} + \sum_{j=1}^{N_{\boldsymbol{X}}} \xi_j \boldsymbol{X}^{(j)} \,. \tag{8.42}$$

The Karhunen-Loève vectors $\boldsymbol{X}^{(j)}$ are of size $N_{\boldsymbol{X}} \times 1$. Extracting from these vectors all elements of $\boldsymbol{X}^{(j)}$ corresponding to a certain $t_k = k\Delta t$ and denoting these vectors by $\boldsymbol{X}^{(j,k)}$ (size $m \times 1$), the corresponding Karhunen-Loève representations of (8.33) and (8.34) read

$$\boldsymbol{\Gamma}_{\boldsymbol{xx}}[:,:,k] = \sum_{j=1}^{N_{\boldsymbol{X}}} \boldsymbol{X}^{(j,k)}\boldsymbol{X}^{(j,k)T} \tag{8.43}$$

and

$$\boldsymbol{\Gamma}_{\boldsymbol{xx}}[:,:,k,l] = \sum_{j=1}^{N_{\boldsymbol{X}}} \boldsymbol{X}^{(j,k)}\boldsymbol{X}^{(j,l)T} \,, \tag{8.44}$$

respectively. Equation (8.43) deserves special attention. As has been already mentioned, $\boldsymbol{\Gamma}_{\boldsymbol{xx}}[:,:,k]$ is of size $m \times m \times N_{\boldsymbol{x}}$, thus – according to the spectral theorem – a complete basis for $\boldsymbol{\Gamma}_{\boldsymbol{xx}}[:,:,k]$ is spanned by the m eigenvectors $\boldsymbol{x}^{(j,k)}$ of $\boldsymbol{\Gamma}_{\boldsymbol{xx}}[:,:,k]$, and not, as indicated in (8.43), by the $N_{\boldsymbol{X}}$ eigenvectors $\boldsymbol{X}^{(j,k)}$. This observation seems at first sight to be of little use because for the determination of the Karhunen-Loève vectors $\boldsymbol{x}^{(j,k)}$ the covariance matrix $\boldsymbol{\Gamma}_{\boldsymbol{xx}}[:,:,k]$ has to be known. However, it should be noted that the basis of $\boldsymbol{\Gamma}_{\boldsymbol{xx}}[:,:,k]$ is considerably smaller than that of $\boldsymbol{\Gamma}_{\boldsymbol{XX}}$, an observation which is discussed in Sec. 13.6.

8.3 Random Fields

Random fields can be seen as an extension of scalar or vector stochastic processes. In the theory of stochastic processes, the parameter t usually indexes a one-dimensional space (temporal or spatial), while in case of random fields this parameter is associated with multi-dimensional spaces. Thus a random field is defined by, see. e.g. [153],

$$x(\boldsymbol{t}); \quad \boldsymbol{t} \in D \subseteq \mathcal{R}^n, \tag{8.45}$$

where

$$\boldsymbol{t} = \begin{Bmatrix} t_1 \\ t_2 \\ \vdots \\ t_n \end{Bmatrix}. \tag{8.46}$$

Second order characteristics are defined similar to those of stochastic processes, i.e.

$$\mu_x(\boldsymbol{t}) = E\{x(\boldsymbol{t})\} \tag{8.47}$$

and

$$\Gamma_{xx}(\boldsymbol{t}, \boldsymbol{s}) = E\{(x(\boldsymbol{t}) - \mu_x(\boldsymbol{t}))(x(\boldsymbol{s}) - \mu_x(\boldsymbol{s}))\} \,. \tag{8.48}$$

Due to the multi-dimensional space over which a random field is defined, the discretization of continuous random fields is more involved that that of (one-dimensional) stochastic processes. A vast amount of literature is devoted to this special field, see e.g. [21, 70, 95] and their references therein, very often in context with stochastic finite elements, see e.g. [135, 52]. The most commonly used discretization methods are point discretization methods (midpoint method, nodal point method, integration point method), the local averaging method, the interpolation method and the series expansion methods (e.g. basis random variable method). Once a proper discretization method has been selected, the random field can be seen as an indexed set of random variables. This is equivalent to the definition of a discrete scalar stochastic process, where at a fixed t, the stochastic process is a random variable. Thus

$$\boldsymbol{x} = \begin{Bmatrix} x(\boldsymbol{t}_0) \\ x(\boldsymbol{t}_1) \\ \vdots \\ x(\boldsymbol{t}_k) \\ \vdots \\ x(\boldsymbol{t}_n) \end{Bmatrix} , \tag{8.49}$$

where t_k may be defined for equal intervals as

$$\boldsymbol{t}_k = \begin{Bmatrix} k_1 \Delta t_1 \\ k_2 \Delta t_2 \\ \vdots \\ k_n \Delta t_n \end{Bmatrix} . \tag{8.50}$$

The Karhunen-Loève representation of a discrete random field has thus been transformed to that of a scalar stochastic process, so that everything stated in Sec. 8.1 is also valid for discrete (univariate) random fields.

8.4 Comparison Between Discrete and Continuous Karhunen-Loève Expansion

The difficulty when dealing with the Karhunen-Loève expansion is related to the the determination of the eigenvalues and the associated eigenvectors of the covariance kernel. Especially for the continuous Karhunen-Loève expansion, their calculation is quite involved. The continuous Karhunen-Loève expansion of a zero mean, second order process is similar to (8.12), i.e.

$$x(t) = \lim_{N \to \infty} \sum_{i=1}^{N} \xi_i \sqrt{\lambda_i} \phi_i(t), \qquad t \in D \subseteq \mathcal{R}^1 . \tag{8.51}$$

However, the simple algebraic eigenvalue problem as stated in (8.14) is replaced in this case by a so-called homogeneous Fredholm integral equation of the second kind, compare with (3.4) or see e.g. [34],

$$\int_D \Gamma_{xx}(t, s) \phi_i(s) \, \mathrm{d}s = \lambda_i \phi_i(t) , \tag{8.52}$$

where the eigenfunctions ϕ satisfy the condition

$$\int_D \phi_i(t) \phi_j(t) \, \mathrm{d}t = \delta_{ij} \tag{8.53}$$

and the random variables ξ are given by

$$\xi_i = \frac{1}{\sqrt{\lambda_i}} \int_D x(t) \phi_i(t) \mathrm{d}t . \tag{8.54}$$

Again, the orthonormality relation for random variables ξ reads

$$E\{\xi_i \xi_j\} = \delta_{ij} . \tag{8.55}$$

Analytical solutions of the Fredholm integral equation do seldom exist, so several numerical algorithms are reported in the literature for the solution of (8.52) for various covariance kernels. For example, in [49] a Galerkin type procedure is used for this purpose, [65] makes use of the Nyström method and [136, 94] apply a wavelet-Galerkin scheme. The majority of the existing numerical solutions for (8.52) transform the integral eigenvalue problem into an algebraic eigenvalue problem, where the transformation represents an additional approximation and computational burden, respectively. However the eigenpairs $(\lambda_i, \phi_i(t))$ of the discrete version of the Karhunen-Loève expansion are obtained by solving *directly* the algebraic eigenvalue problem (8.14) instead of (8.52), which overcomes the shortcomings of the aforementioned transformation. Nevertheless, the accuracy of the discrete version of the Karhunen-Loève expansion when compared to its continuous form has to be addressed. In particular, the eigenvalues and the associated eigenvectors obtained by the algebraic and the integral eigenvalue problem, respectively, are compared. A covariance kernel commonly used in stochastic finite elements is, see Fig. 8.1,

$$\Gamma_{xx}(t, s) = \mathrm{e}^{-|t-s|/b} , \qquad t \in (-a, a) . \tag{8.56}$$

In this case, an analytical solution of the eigenfunctions of (8.53) is available,

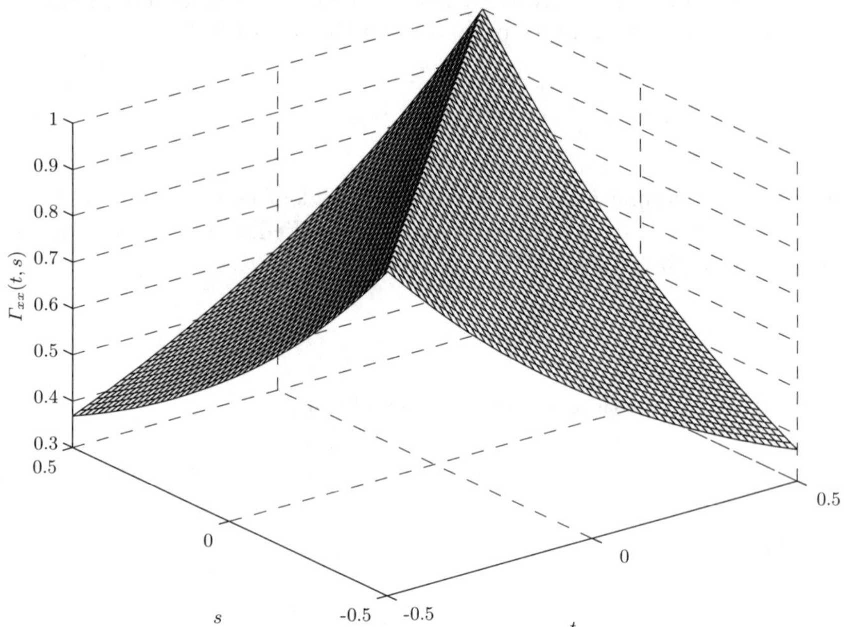

Fig. 8.1. Surface plot of covariance function $\Gamma_{xx}(t,s) = e^{-|t-s|/b}$ in the interval [-0.5,0,5], $b = 1$.

see e.g. [49]. The eigenpairs for even i are given by

$$\lambda_i = \frac{2c}{\omega_i^2 + c^2} , \qquad \phi_i(t) = \frac{\cos(\omega_i t)}{\sqrt{a + \frac{\sin(2\omega_i a)}{2\omega_i}}} , \tag{8.57}$$

where

$$c - \omega \tan(\omega a) = 0 . \tag{8.58}$$

Similar, for odd i, the eigenpairs $(\lambda_i, \phi_i(t))$ are defined by

$$\lambda_i = \frac{2c}{\omega_i^2 + c^2} , \qquad \phi_i(t) = \frac{\sin(\omega_i t)}{\sqrt{a - \frac{\sin(2\omega_i a)}{2\omega_i}}} , \tag{8.59}$$

where

$$\omega + c \tan(\omega a) = 0 . \tag{8.60}$$

As can be seen from Figs. 8.2–8.3 (compare also to Figs. 2.1–2.2 in [49]), the discrete form of the Karhunen-Loève expansion agrees very well with the analytical solution of its continuous counterpart. It has to be noted that the eigenfunctions for the discrete Karhunen-Loève expansion in Fig. 8.2 are normalized by (8.53) instead of (8.15). The eigenvalues have been weighted accordingly.

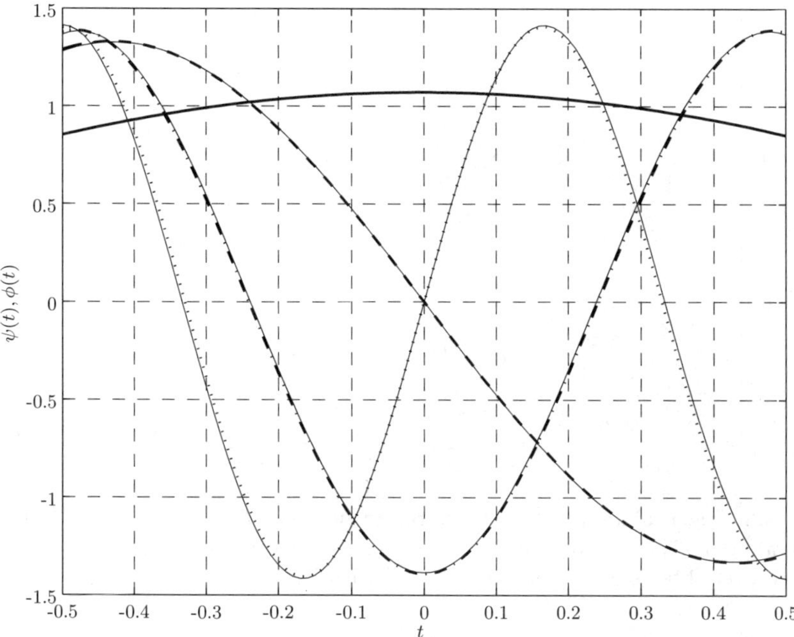

Fig. 8.2. Eigenfunctions $\{\psi_i(t)\}_{i=1}^4$ of covariance function $\Gamma_{xx}(t,s) = e^{-|t-s|/b}$ in the interval [-0.5,0,5], $b = 1$ obtained by (8.14): $i = 1$ (solid), $i = 2$ (dashed), $i = 3$ (dash-dot), $i = 4$ (dotted). Corresponding analytical eigenfunctions $\phi(t)$ obtained by (8.52) are plotted with a thin solid line.

The results from this simple test case can be generalized and allow to summarize some important features of the discrete Karhunen-Loève expansion:

(i) The solution of the algebraic eigenvalue problem as given by (8.14) is available as a standard procedure in many software packages, i.e. sophisticated algorithms for solving efficiently the standard eigenvalue problem of symmetric matrices can be applied.

(ii) Contrary to many numerical solutions of (8.53), the eigenvalue problem can be applied directly on the covariance matrix, i.e. no prior transformation of the integral eigenvalue problem is required. It should be mentioned, that numerical solutions of (8.53) which are based on a series expansion of the eigenvectors of the covariance kernel, introduce an additional numerical approximation.

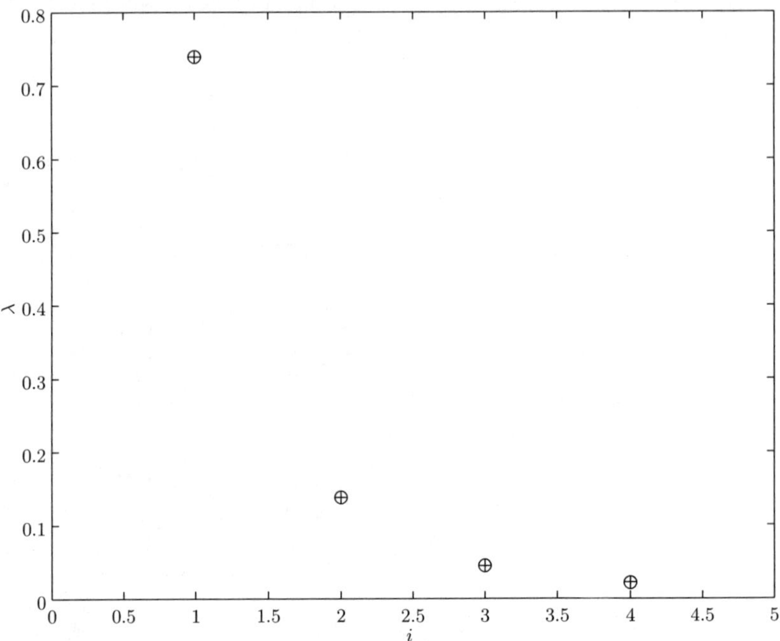

Fig. 8.3. Eigenvalues $\{\lambda_i\}_{i=1}^4$ of covariance function $\Gamma_{xx}(t,s) = e^{-|t-s|/b}$ in the interval [-0.5,0,5], $b = 1$ obtained by (8.14) (circle). Corresponding analytical eigenvalues $\lambda_i(t)$ obtained by (8.52) are denoted with "+".

(iii) The eigenpairs obtained by the discrete Karhunen-Loève expansion agree very well with their continuous counterparts, being theoretically identical if an infinite small interval for discretization is used.

9

Direct Monte Carlo Simulation

The evaluation of the stochastic response by the Monte Carlo simulation technique is a powerful method for highly nonlinear systems as well as for systems where the input is modeled by large number of random variables.

9.1 Basic Principles

Figure 9.1 depicts the basic principles of Monte Carlo sampling, where the laws of statistics are exploited to derive information on the variability of the response. By using a suitable number generator (see e.g. randlib [23]) statistically independent samples of the input are generated by a type of game of chance, and which follow the prescribed probability distributions of the uncertain parameters. Let the system be described by the operator L, so that a set of random input variables collected in a vector $\boldsymbol{x}$ defined in a m-dimensional

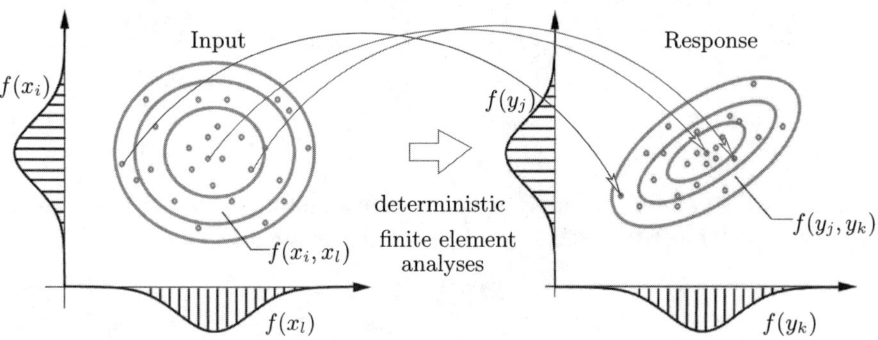

Fig. 9.1. Stochastic Analysis based on Monte Carlo Sampling

vector space is mapped to the r-dimensional output $\boldsymbol{y}$, i.e.

$$L\boldsymbol{x} = \boldsymbol{y} \ . \tag{9.1}$$

In the simplest form of the Monte Carlo simulation technique – denoted as direct Monte Carlo simulation – for each generated sample of the input $\boldsymbol{x}^{(i)}$ the corresponding output $\boldsymbol{y}^{(i)}$ is calculated. Hence, the input distribution $f(x_1, x_2, \ldots, x_m)$ is represented according to statistical laws by a finite number n of independent samples $\{\boldsymbol{x}^{(i)}\}_{i=1}^n$. Each vector $\boldsymbol{x}^{(i)}$ specifies for each uncertain parameter a deterministic discrete value and consequently defines deterministically the response which might be represented by the vector

$$\boldsymbol{y}^{(i)} = L\boldsymbol{x}^{(i)} \ . \tag{9.2}$$

Hence, traditional deterministic finite element analyses can be used to provide the mapping given by (9.2) between input and response.
In the simplest case, it might be justified to assume that all uncertainties are independent. Such an assumption is reasonable as long as this assumption does not contradict experience and physical properties. When the components are considered as independent, each component can be generated by available random number generators where distribution and its parameters must be supplied. The above simple and straight forward procedure is getting more involved for cases where correlations between random variables need to be considered, as it is the case for stochastic processes or random fields.

9.2 Error Assessment

Let the unknown mean and variance of a component y_k of the output be denoted with

$$\mu_{y_k} = E\{y_k\} \qquad \text{and} \qquad \sigma_{y_k}^2 = E\{(y_k - \mu_{y_k})^2\} \ , \tag{9.3}$$

respectively. The estimator of the response is defined by

$$\bar{y}_k = \frac{1}{n} \sum_{i=1}^{n} y_k^{(i)} \ . \tag{9.4}$$

and is itself a random variable. It can be shown, see e.g. [45], that the mean and the variance of the estimator of the response are given by

$$E\{\bar{y}_k\} = \mu_{y_k} \tag{9.5}$$

and

$$\mathrm{Var}(\bar{y}_k) = E\{(\bar{y}_k - E\{\bar{y}_k\})^2\} = \frac{\sigma_{y_k}^2}{n} \ . \tag{9.6}$$

Tschebyschev's inequality provides a basis for error assessment, i.e.

$$P\{|\bar{y} - \mu_y| < \epsilon\} \geq 1 - \frac{1}{\epsilon^2}\frac{\sigma_y^2}{n} , \tag{9.7}$$

where ϵ denotes a tolerance. By specifying a confidence level $1 - \delta$, a "worst case" minimum sample size can be calculated by

$$n_{\min} \geq \frac{\sigma_{y_k}^2}{\epsilon^2}\frac{1}{\delta} . \tag{9.8}$$

By definition, (9.8) is valid for any sample size and any probability distribution of y_k and specifies a larger sample size than is necessary. Due to the central limit theorem, however, the sum of a large number of not necessarily Gaussian random variables yields a Gaussian random variable. Specifically, the distribution of $\bar{y}_k$ is $N(\mu_{y_k}, \sigma_{y_k}/\sqrt{n})$, so that for $n \to \infty$ it follows that $\bar{y}_k \to \mu_{y_k}$. This allow one to specify a minimum sample size for a sufficiently large number n of simulations under the assumption that $\sigma_{y_k}^2 \approx \mathrm{Var}(y_k)$ ("known variance")

$$n_{\mathrm{N},\min} \geq \frac{\sigma_y^2}{\epsilon^2}\left[\Phi^{-1}(1 - \delta/2)\right]^2 , \tag{9.9}$$

where $\Phi^{-1}(\cdot)$ denotes the inverse of the normal cumulative distribution function. Equivalently, for a given number n of simulations and a specified confidence level, the corresponding confidence interval for μ_{y_k} is given by

$$\mu_{y_k,1-\delta} = \left(\hat{y_k} - \Phi^{-1}(1 - \delta/2)\frac{\sigma_{y_k}}{\sqrt{n}} ; \hat{y_k} + \Phi^{-1}(1 - \delta/2)\frac{\sigma_{y_k}}{\sqrt{n}}\right) . \tag{9.10}$$

Among all methods that utilize a n-point estimation in the m-dimensional parameter space, the direct Monte Carlo method has an absolute estimation error that decreases with $n^{-1/2}$, independently of the dimension m as seen from (9.6), whereas all other approaches have errors which decrease with $n^{-1/m}$ at best, see e.g. [45].

10

Equivalent Statistical Linearization

The solution of non-linear vibration problems in terms of probabilistic characteristics of the response is a complex task, since no general methods of solution are available. If second order statistics of the response have to be calculated, however, the method of equivalent statistical linearization , see e.g. [109, 140], is conceptional simple and straight forward to apply.
In the following, the concept of equivalent statistical linearization is first reviewed for scalar valued transformations and extended later on for vector valued transformations, which commonly appear in non-linear finite element models.

10.1 Scalar Valued Transformations

A non-linear scalar valued function representing the rate of change of the process $z(t)$ can be seen as a transformation with memory of the input $\boldsymbol{y}(t)$ to the output $z(t)$, i.e.

$$g(z(t), \boldsymbol{y}(t)) = \dot{z}(t) \;, \tag{10.1}$$

where

$$\boldsymbol{y}(t) = \begin{Bmatrix} y_1(t) \\ y_2(t) \\ \vdots \\ y_n(t) \end{Bmatrix} . \tag{10.2}$$

The non-linear scalar valued function is replaced by a linear one

$$\begin{aligned} a_1(t)\tilde{y}_1(t) + a_2(t)\tilde{y}_2(t) + \ldots + a_n(t)\tilde{y}_n(t) + b(t)\tilde{z}(t) + c(t) \\ = \boldsymbol{a}(t)^T\tilde{\boldsymbol{y}}(t) + b(t)\tilde{z}(t) + c(t) = \dot{z}(t) \;, \end{aligned} \tag{10.3}$$

where $\tilde{\boldsymbol{y}}(t)$ denotes the centered input process

$$\tilde{\boldsymbol{y}}(t) = \boldsymbol{y}(t) - \boldsymbol{\mu}_{\boldsymbol{y}}(t) \;, \tag{10.4}$$

$\tilde{z}(t) = z(t) - \mu_z(t)$ and the vector

$$\boldsymbol{a}(t) = \begin{Bmatrix} a_1(t) \\ a_2(t) \\ \vdots \\ a_n(t) \end{Bmatrix} \tag{10.5}$$

comprises the so-called linearization coefficients, respectively. By defining the error between the non-linear and the linear system by

$$\epsilon(t) = g(z(t), \boldsymbol{y}(t)) - \boldsymbol{a}(t)^T \tilde{\boldsymbol{y}}(\boldsymbol{t}) - b(t)\tilde{z}(t) - c(t) , \tag{10.6}$$

the equivalent statistical linearization technique minimizes this error in the mean square sense, i.e.

$$e(t) = E\{\epsilon^2(t)\} \rightarrow \min , \tag{10.7}$$

where the minimum can be obtained from the conditions

$$\frac{\partial e(t)}{\partial a_i(t)} = 0, \qquad \frac{\partial e(t)}{\partial b(t)} = 0, \qquad \text{and} \qquad \frac{\partial e(t)}{\partial c(t)} = 0 . \tag{10.8}$$

By direct calculations one arrives at the set of algebraic equations

$$E\{\tilde{\boldsymbol{y}}(t)\tilde{\boldsymbol{y}}(t)^T\}\boldsymbol{a}(t) = E\{\tilde{\boldsymbol{y}}(t)g(z(t), \boldsymbol{y}(t))\} , \tag{10.9}$$

$$E\{\tilde{z}(t)\tilde{z}(t)\}b(t) = E\{\tilde{z}(t)g(z(t), \boldsymbol{y}(t))\} , \tag{10.10}$$

and

$$c(t) = E\{g(z(t), \boldsymbol{y}(t))\} . \tag{10.11}$$

For a Gaussian input process it has been shown in [9], that the relation

$$E\{\tilde{\boldsymbol{y}}(t)g(z(t), \boldsymbol{y}(t))\} = E\{\tilde{\boldsymbol{y}}(\boldsymbol{t})\tilde{\boldsymbol{y}}(\boldsymbol{t})^T\}E\{\boldsymbol{\nabla} g(z(t), \boldsymbol{y}(t))\} \tag{10.12}$$

holds, where

$$\boldsymbol{\nabla} = \begin{Bmatrix} \frac{\partial}{\partial y_1} \\ \frac{\partial}{\partial y_2} \\ \vdots \\ \frac{\partial}{\partial y_n} \end{Bmatrix} , \tag{10.13}$$

and thus (10.9) can be simplified to

$$\boldsymbol{a}(t) = E\{\boldsymbol{\nabla} g(z(t), \boldsymbol{y}(t))\} \qquad \text{or} \qquad a_i(t) = E\left\{\frac{\partial g(z(t), \boldsymbol{y}(t))}{\partial y_i}\right\} . \tag{10.14}$$

In a similar way, also a relation for $b(t)$ can be derived.

10.2 Vector Valued Transformations

In a finite element analysis, where a hysteretic behavior of the system is generally modeled by several non-linear elements, the concept reviewed in section 10.1 has to be extended to vector valued transformation. In this case, the non-linear relation reads

$$\boldsymbol{g}(\boldsymbol{z}(t), \boldsymbol{y}(t)) = \dot{\boldsymbol{z}}(t) , \tag{10.15}$$

where

$$\boldsymbol{g}(\boldsymbol{z}(t), \boldsymbol{y}(t)) = \begin{Bmatrix} g_1(\boldsymbol{z}(t), \boldsymbol{y}(t)) \\ g_2(\boldsymbol{z}(t), \boldsymbol{y}(t)) \\ \vdots \\ g_m(\boldsymbol{z}(t), \boldsymbol{y}(t)) \end{Bmatrix} . \tag{10.16}$$

Again, this transformation is replaced by a linear one,

$$\boldsymbol{A}(t)\tilde{\boldsymbol{y}}(t) + \boldsymbol{B}(t)\tilde{\boldsymbol{z}}(t) + \boldsymbol{c}(t) = \dot{\boldsymbol{z}}(t) , \tag{10.17}$$

where the linearization coefficients are now elements of the $m \times n$ matrix $\boldsymbol{A}(t)$ and the $m \times m$ matrix $\boldsymbol{B}(t)$. Applying the same steps as shown previously for scalar valued transformations, one arrives at

$$a_{ij} = E\left\{ \frac{\partial g_i(\boldsymbol{z}(t), \boldsymbol{y}(t))}{\partial y_j} \right\} , \tag{10.18}$$

$$b_{ij} = E\left\{ \frac{\partial g_i(\boldsymbol{z}(t), \boldsymbol{y}(t))}{\partial z_j} \right\} , \tag{10.19}$$

and

$$\boldsymbol{c} = E\left\{ \boldsymbol{g}(\boldsymbol{z}(t), \boldsymbol{y}(t)) \right\} . \tag{10.20}$$

11

Random Vibrations of Large Finite Element Systems

The basic idea for the computation of the stochastic response is based on the Karhunen-Loève expansion of the stochastic loading. For each Karhunen-Loève vector of the loading the corresponding Karhunen-Loève vector of the response of the system is computed. This allows – applying the superposition principle – to calculate the covariance matrix of the output quantities (e.g. displacements, velocities etc.).

Quite frequently, however, only second moment characteristics of a few degrees of freedom are of interest, which can also be determined most efficiently by the described procedure without calculating the full covariance matrix. This feature is especially important for systems modeled with a large number n of degrees of freedom, since the storage of the non-sparse covariance matrix with dimensions $2n \times 2n$ for large systems represents still nowadays a hardware problem.

In presence of non-linearities, the system equations are linearized by the equivalent statistical linearization technique, so that the superposition principle is still valid. The described procedure is straightforward and attractive for the deterministic engineering community, since deterministic integration algorithms as part of many commercial finite element packages are applied for the integration of Karhunen-Loève vectors. The efficiency of the described method can be significantly enhanced when the deterministic integration is carried out in a reduced subspace, i.e. by applying a mode superposition technique.

11.1 Stochastic Excitation

The external loading $\boldsymbol{f}(t)$ as used in (6.4) or (6.6) is assumed to be a simplified representation of earthquake excitation, i.e.

$$\boldsymbol{f}(t) = -\boldsymbol{M}\boldsymbol{I}_{\boldsymbol{a}}\boldsymbol{a}(t) \,, \tag{11.1}$$

where $\boldsymbol{I_a}$ is a displacement transformation matrix that expresses the displacement of each degree of freedom due to static application of a unit support displacement, see e.g. [33]. Vector $\boldsymbol{a}(t)$ denotes the discrete, stochastic earthquake acceleration

$$\boldsymbol{a}(t) = \begin{Bmatrix} a_x(t) \\ a_y(t) \\ a_z(t) \end{Bmatrix} \tag{11.2}$$

applied at the structure's supports. The components $a_x(t)$, $a_y(t)$ and $a_z(t)$, are assumed to be mutually statistically independent. The Karhunen-Loève representation of e.g. $a_x(t)$, can be written as, see (8.12),

$$a_x(t) = a_x^{(0)}(t) + \sum_{j=1}^{N_a} \xi_{j_x} a_x^{(j)}(t) \approx a_x^{(0)}(t) + \sum_{j=1}^{m} \xi_{j_x} a_x^{(j)}(t) \ , \tag{11.3}$$

where $a_x^{(0)}(t)$ and $a_x^{(j)}(t)$ denote the mean function and the j-th Karhunen-Loève vector of $a_x(t)$, respectively, and m is the order of truncation of the series expansion. Thus, the truncated Karhunen-Loève representation of $\boldsymbol{a}(t)$ is given by

$$\boldsymbol{a}(t) \approx \boldsymbol{a}^{(0)}(t) + \sum_{j=1}^{m} \boldsymbol{\xi}_j \boldsymbol{a}^{(j)}(t) \ , \tag{11.4}$$

where

$$\boldsymbol{\xi}_j = \begin{bmatrix} \xi_{j_x} & 0 & 0 \\ 0 & \xi_{j_y} & 0 \\ 0 & 0 & \xi_{j_z} \end{bmatrix} . \tag{11.5}$$

In case the components of $\boldsymbol{a}(t)$ are correlated, a Karhunen-Loève representation in the form of (8.42), see Sec. 8.2, would be necessary. For uncorrelated ground accelerations, however, the Karhunen-Loève representation of the loading vector is given by

$$\begin{aligned} \boldsymbol{f}(t) &\approx \boldsymbol{f}^{(0)}(t) + \sum_{j=1}^{3m} \bar{\xi}_j \boldsymbol{f}^{(j)}(t) \\ &\approx -\boldsymbol{M}\boldsymbol{I}_a\Big(\boldsymbol{a}^{(0)}(t) + \sum_{j=1}^{m} \boldsymbol{\xi}_j \boldsymbol{a}^{(j)}(t)\Big) \ , \end{aligned} \tag{11.6}$$

where, again, $\boldsymbol{f}^{(0)}(t)$ and $\boldsymbol{f}^{(j)}(t)$ denote the mean function and the j-th Karhunen-Loève vector of $\boldsymbol{f}(t)$, respectively, and

$$\begin{aligned} \{\bar{\xi}_j\}_{j=1}^{m} &= \{\xi_{j_x}\}_{j=1}^{m} \ , \\ \{\bar{\xi}_j\}_{j=m+1}^{2m} &= \{\xi_{j_y}\}_{j=1}^{m} \ , \\ \{\bar{\xi}_j\}_{j=2m+1}^{3m} &= \{\xi_{j_z}\}_{j=1}^{m} \ . \end{aligned} \tag{11.7}$$

11.2 Response Statistics

Based on the superposition principle and the representation of the loading as a sum of independent loading vectors according to (11.6), the structural response in terms of displacements, velocities and auxiliary variables, see (6.4),

$$\boldsymbol{x}(t) = \{\boldsymbol{u}^T(t), \dot{\boldsymbol{u}}^T(t), \boldsymbol{q}^T(t)\}^T , \tag{11.8}$$

can be represented by

$$\boldsymbol{x}(t) = \boldsymbol{x}^{(0)}(t) + \sum_{j=1}^{3m} \bar{\xi}_j \boldsymbol{x}^{(j)}(t) . \tag{11.9}$$

Now all vectors $\{\boldsymbol{x}^{(j)}\}_{j=0}^{3m}$, which uniquely specify the first two moments of the stochastic response

$$\boldsymbol{\mu}_{\boldsymbol{x}}(t) = \boldsymbol{x}^{(0)}(t) \tag{11.10}$$

and

$$\begin{aligned} \boldsymbol{\Gamma}_{\boldsymbol{x}\boldsymbol{x}}(t,s) &= E\{(\boldsymbol{x}(t) - \boldsymbol{\mu}_x(t))(\boldsymbol{x}(s) - \boldsymbol{\mu}_x(s))^T\} \\ &= \sum_{j=1}^{3m} \boldsymbol{x}^{(j)}(t)\boldsymbol{x}^{(j)T}(s) \end{aligned} \tag{11.11}$$

have to be determined. For each deterministic Karhunen-Loève vector of the loading the associated deterministic structural response can be calculated. Symbolically, this can be written as

$$\{\boldsymbol{f}^{(j)}\}_{j=0}^{3m} \mapsto \left(\{\boldsymbol{u}^{(j)}\}_{j=0}^{3m}, \{\dot{\boldsymbol{u}}^{(j)}\}_{j=0}^{3m}, \{\boldsymbol{q}^{(j)}\}_{j=0}^{3m}\right) . \tag{11.12}$$

The second moment characteristics of the structural response can be calculated according to

$$\begin{aligned} \boldsymbol{\mu}_{\boldsymbol{u}}(t) &= E\{\boldsymbol{u}(t)\} = \boldsymbol{u}^{(0)}(t) , \\ \boldsymbol{\mu}_{\dot{\boldsymbol{u}}}(t) &= E\{\dot{\boldsymbol{u}}(t)\} = \dot{\boldsymbol{u}}^{(0)}(t) , \\ \boldsymbol{\mu}_{\boldsymbol{q}}(t) &= E\{\boldsymbol{q}(t)\} = \boldsymbol{q}^{(0)}(t) , \end{aligned} \tag{11.13}$$

and

$$\begin{aligned} \boldsymbol{\Gamma}_{\boldsymbol{u}\boldsymbol{u}}(t,s) &= \sum_{j=1}^{3m} \boldsymbol{u}^{(j)}(t)\boldsymbol{u}^{(j)^T}(s) , \\ \boldsymbol{\Gamma}_{\dot{\boldsymbol{u}}\dot{\boldsymbol{u}}}(t,s) &= \sum_{j=1}^{3m} \dot{\boldsymbol{u}}^{(j)}(t)\dot{\boldsymbol{u}}^{(j)^T}(s) , \\ \boldsymbol{\Gamma}_{\boldsymbol{q}\boldsymbol{q}}(t,s) &= \sum_{j=1}^{3m} \boldsymbol{q}^{(j)}(t)\boldsymbol{q}^{(j)^T}(s) . \end{aligned} \tag{11.14}$$

It remains now to show how the Karhunen-Loève vectors of the structural response quantities – depending on the type of system – can be calculated.

11.2.1 Linear Systems

For linear systems described by (6.6), response statistics can be calculated in a straightforward manner. Using direct integration, this requires the solution of

$$\boldsymbol{M}\,\ddot{\boldsymbol{u}}^{(j)} + \boldsymbol{C}\,\dot{\boldsymbol{u}}^{(j)} + \boldsymbol{K}\,\boldsymbol{u}^{(j)} = \boldsymbol{f}^{(j)} \qquad j \geq 0\,, \tag{11.15}$$

which is also valid for non-zero mean problems. When applying a mode superposition method like e.g. the mode acceleration method, see Sec. 6.4, the set of equations given by

$$\begin{gathered}
\ddot{z}_i^{(j)} + \dot{z}_i^{(j)} 2\zeta_i\omega_i + z_i^{(j)}\omega_i^2 = \boldsymbol{\phi}_i^T\,\boldsymbol{f}^{(j)}\,,\\
\boldsymbol{u}_d^{(j,s)} = \boldsymbol{\Phi}\,\boldsymbol{z}^{(j,s)}\,, \qquad \dot{\boldsymbol{u}}_d^{(j,s)} = \boldsymbol{\Phi}\,\dot{\boldsymbol{z}}^{(j,s)}\,,\\
\boldsymbol{u}_s^{(j,s)} = \hat{\boldsymbol{B}}\,\boldsymbol{a}^{(j)}\,, \qquad \dot{\boldsymbol{u}}_s^{(j,s)} = \hat{\boldsymbol{B}}\,\dot{\boldsymbol{a}}^{(j)}\,,\\
\boldsymbol{u}^{(j,s)} = \boldsymbol{u}_d^{(j,s)} + \boldsymbol{u}_s^{(j,s)}\,,\\
\dot{\boldsymbol{u}}^{(j,s)} = \dot{\boldsymbol{u}}_d^{(j,s)} + \dot{\boldsymbol{u}}_s^{(j,s)}
\end{gathered} \tag{11.16}$$

has to be solved.

11.2.2 Non-Linear Systems

The non-linearity treated in the following is of hysteretic type. A simplified hysteretic behavior can be described by non-linear ordinary differential equations, which are based on the well known Bouc-Wen model, see e.g. [157]. The constants appearing in these equations may be adapted such that a wide variety of hysteresis loops may be modeled. The hysteretic elements described in section 13.2 can be seen as non-linear transformations with memory, since the auxiliary variables $\boldsymbol{q}(t)$ at an arbitrary time t, representing the output of the transformation, depend on the entire history of the system response defined by $\boldsymbol{u}(t)$ and $\dot{\boldsymbol{u}}(t)$.

Modeling in Local Element Coordinates

The most suitable and straightforward approach in modeling the non-linear restoring force $\bar{\boldsymbol{r}}(\boldsymbol{u}, \dot{\boldsymbol{u}}, \boldsymbol{q})$, see (6.4), is in local element coordinates [43, 100]. Hereby, $\bar{\boldsymbol{r}}(\boldsymbol{u}, \dot{\boldsymbol{u}}, \boldsymbol{q})$ is represented in a minimal number of coordinates, where in the following the superscript "$^{(e)}$" denotes the non-linear element and the subscript "$_L$" indicates local coordinates. For global coordinates the subscript "$_G$" is omitted where the meaning is clear.
The local element displacements $\boldsymbol{u}_L^{(e)}$ and velocities $\dot{\boldsymbol{u}}_L^{(e)}$ are related to the global nodal displacements $\boldsymbol{u}^{(e)}$ and velocities $\dot{\boldsymbol{u}}^{(e)}$, respectively, by a linear transformation $\boldsymbol{T}^{(e)}$, i.e.

$$\boldsymbol{u}_L^{(e)} = \boldsymbol{T}^{(e)}\boldsymbol{u}^{(e)}\,, \qquad \dot{\boldsymbol{u}}_L^{(e)} = \boldsymbol{T}^{(e)}\dot{\boldsymbol{u}}^{(e)}\,, \tag{11.17}$$

where

$$\boldsymbol{u}^{(e)} = \begin{Bmatrix} \boldsymbol{u}_1^{(e)} \\ \boldsymbol{u}_2^{(e)} \end{Bmatrix} \tag{11.18}$$

is a vector of the global nodal degrees of freedom to which the non-linear element is related. The element specific quantity $\boldsymbol{q}^{(e)}$ does not require a distinction between local and global, i.e.

$$\boldsymbol{q}_L^{(e)} = \boldsymbol{q}^{(e)} . \tag{11.19}$$

The non-linear part of $\boldsymbol{r}_L^{(e)}$, i.e. $\bar{\boldsymbol{r}}_L^{(e)}$, can thus be calculated by, see (6.5),

$$\bar{\boldsymbol{r}}_L^{(e)} = -\boldsymbol{R}_L^{(e)} \boldsymbol{q}^{(e)}(\boldsymbol{u}_L^{(e)}, \dot{\boldsymbol{u}}_L^{(e)}) . \tag{11.20}$$

Using the fundamental transformation laws, the loading in global coordinates can be determined by

$$\bar{\boldsymbol{r}}_G^{(e)} = \boldsymbol{T}^{(e)T} \bar{\boldsymbol{r}}_L^{(e)} = -\boldsymbol{R}_G^{(e)} \boldsymbol{q}^{(e)}(\boldsymbol{u}_L^{(e)}, \dot{\boldsymbol{u}}_L^{(e)}) , \tag{11.21}$$

where

$$\boldsymbol{R}_G^{(e)} = \boldsymbol{T}^{(e)T} \boldsymbol{R}_L^{(e)} . \tag{11.22}$$

Assembling all element matrices $\boldsymbol{R}_G^{(e)}$ and $\boldsymbol{q}^{(e)}$ into a matrix and a vector, respectively,

$$\sum_e \boldsymbol{R}_G^{(e)} \mapsto \boldsymbol{R} , \qquad \sum_e \boldsymbol{q}^{(e)} \mapsto \boldsymbol{q} , \tag{11.23}$$

allows one to represent the additional loading due to the non-linear elements by the product

$$\bar{\boldsymbol{r}}(\boldsymbol{u}, \dot{\boldsymbol{u}}, \boldsymbol{q}) = -\boldsymbol{R}\boldsymbol{q}(\boldsymbol{u}, \dot{\boldsymbol{u}}) . \tag{11.24}$$

Equivalent Statistical Linearization

The described procedure relies on the Karhunen-Loève representation of the covariance matrix of the response. As it is well known that the superposition principle is valid only for linear systems described by (6.6), the non-linear system equations given by (6.4) have to be linearized for each time step. For this purpose equivalent statistical linearization is applied, see Sec. 10, which in fact has been developed from the deterministic equivalent linearization procedure.

Applying equivalent statistical linearization to general non-linear elements implies integration over an r-dimensional space, where r denotes the length of the state vector required to specify the non-linear element. An accurate integration over larger dimensions, say $r > 4$, is computationally not feasible. An approximation using Monte Carlo simulation in combination with a least square solution has been suggested as a generally applicable solution in

[99, 121]. The described procedure, however, uses equivalent statistical linearization as an existing tool in stochastic mechanics and thus in this monograph the well tested one-dimensional non-linear elements have been used for modeling the non-linear system behavior. Linearizing the non-linear vector function (6.2) leads to a linear first order vector differential equation for the temporal evolution of the auxiliary variables, see Chap. 10, i.e.

$$\dot{\boldsymbol{q}} = \boldsymbol{g}(\boldsymbol{u}, \dot{\boldsymbol{u}}, \boldsymbol{q}) \approx \boldsymbol{b} + \boldsymbol{U}\tilde{\boldsymbol{u}} + \boldsymbol{V}\dot{\tilde{\boldsymbol{u}}} + \boldsymbol{W}\tilde{\boldsymbol{q}} \,, \tag{11.25}$$

where the vector $\boldsymbol{b}$ and the matrices $\boldsymbol{U}$,$\boldsymbol{V}$ and $\boldsymbol{W}$ contain the so-called linearization coefficients with

$$\boldsymbol{b} = E\{\boldsymbol{g}(\boldsymbol{u}, \dot{\boldsymbol{u}}, \boldsymbol{q})\} \,, \tag{11.26}$$

and zero mean vectors

$$\begin{aligned} \tilde{\boldsymbol{u}} &= \boldsymbol{u} - \boldsymbol{\mu}_{\boldsymbol{u}} \,, \quad \boldsymbol{\mu}_{\boldsymbol{u}} = E\{\boldsymbol{u}\} \,, \\ \dot{\tilde{\boldsymbol{u}}} &= \dot{\boldsymbol{u}} - \boldsymbol{\mu}_{\dot{\boldsymbol{u}}} \,, \quad \boldsymbol{\mu}_{\dot{\boldsymbol{u}}} = E\{\dot{\boldsymbol{u}}\} \,, \\ \tilde{\boldsymbol{q}} &= \boldsymbol{q} - \boldsymbol{\mu}_{\boldsymbol{q}} \,, \quad \boldsymbol{\mu}_{\boldsymbol{q}} = E\{\boldsymbol{q}\} \,. \end{aligned} \tag{11.27}$$

In equivalent statistical linearization, the input and consequently the response are commonly assumed to be jointly Gaussian distributed, whereby second order characteristics of the response have to be known in advance. On the other hand, the response statistics depend on the linearization coefficients contained in $\boldsymbol{b}$, $\boldsymbol{U}$,$\boldsymbol{V}$ and $\boldsymbol{W}$ as determined by equivalent statistical linearization. This mutual dependency is resolved by an iterative procedure, as will be described in detail in the next section.

Direct Integration

The following is a detailed description of an iteration scheme for the calculation of the unknown structural response as written symbolically in (11.12). Hereby the zero mean loading process as well as a symmetric hysteresis loop is assumed. Non-zero mean problems are treated in Sec. 11.3. It is assumed that the system response is known for time t, i.e.

$$^{t}\{\boldsymbol{u}^{(j)}\}_{j=0}^{3m} \,, \qquad ^{t}\{\dot{\boldsymbol{u}}^{(j)}\}_{j=0}^{3m} \qquad \text{and} \qquad ^{t}\{\boldsymbol{q}^{(j)}\}_{j=0}^{3m} \tag{11.28}$$

and is sought for time $t + \Delta t$, i.e.

$$^{t+\Delta t}\{\boldsymbol{u}^{(j)}\}_{j=0}^{3m} \,, \qquad ^{t+\Delta t}\{\dot{\boldsymbol{u}}^{(j)}\}_{j=0}^{3m} \qquad \text{and} \qquad ^{t+\Delta t}\{\boldsymbol{q}^{(j)}\}_{j=0}^{3m} \,. \tag{11.29}$$

Using direct integration, (6.4) has to be solved for each deterministic Karhunen-Loève vector according to

$$\begin{aligned} \boldsymbol{M}\,^{t+\Delta t}\ddot{\boldsymbol{u}}^{(j,s)} + \boldsymbol{C}\,^{t+\Delta t}\dot{\boldsymbol{u}}^{(j,s)} + \boldsymbol{K}\,^{t+\Delta t}\boldsymbol{u}^{(j,s)} = \\ ^{t+\Delta t}\boldsymbol{f}^{(j)} + \boldsymbol{R}\,^{t+\Delta t}\boldsymbol{q}^{(j,s-1)} \,, \end{aligned} \tag{11.30}$$

where s denotes the iteration cycle. Due to the additional loading obtained by the shift of the non-linear restoring force vector to the right hand site of (6.4), the Karhunen-Loève vectors can not be calculated directly, but in an iterative manner. At the beginning of each iteration it is assumed that the relation

$$^{t+\Delta t}\boldsymbol{q}^{(j,0)} := {}^{t}\boldsymbol{q}^{(j)} , \tag{11.31}$$

holds. The linearization of the non-linear elements is done independently in local coordinates. Therefore, the set of equations (11.25)–(11.27) is rewritten in local coordinates, yielding

$$\dot{\boldsymbol{q}}^{(e,s)} \approx \boldsymbol{b}^{(e,s)} + \boldsymbol{U}^{(e,s)}\tilde{\boldsymbol{u}}_L^{(e,s)} + \boldsymbol{V}^{(e,s)}\dot{\tilde{\boldsymbol{u}}}_L^{(e,s)} + \boldsymbol{W}^{(e,s)}\tilde{\boldsymbol{q}}^{(e,s-1)} , \tag{11.32}$$

with

$$\boldsymbol{b}^{(e,s)} = E\{\boldsymbol{g}^{(e,s)}(\boldsymbol{u}_L^{(e,s)}, \dot{\boldsymbol{u}}_L^{(e,s)}, \boldsymbol{q}^{(e,s-1)})\} \tag{11.33}$$

and the zero mean processes

$$\begin{aligned} \tilde{\boldsymbol{u}}_L^{(e)} &= \boldsymbol{u}_L^{(e)} - \boldsymbol{\mu}_{\boldsymbol{u}_L^{(e)}} , \\ \dot{\tilde{\boldsymbol{u}}}_L^{(e)} &= \dot{\boldsymbol{u}}_L^{(e)} - \boldsymbol{\mu}_{\dot{\boldsymbol{u}}_L^{(e)}} , \\ \tilde{\boldsymbol{q}}^{(e)} &= \boldsymbol{q}^{(e)} - \boldsymbol{\mu}_{\boldsymbol{q}^{(e)}} . \end{aligned} \tag{11.34}$$

The linearization in local coordinates requires the first two moments of the stochastic response

$$\boldsymbol{x}^{(e,s)} = \begin{Bmatrix} \boldsymbol{u}_L^{(e,s)} \\ \dot{\boldsymbol{u}}_L^{(e,s)} \\ \boldsymbol{q}^{(e,s-1)} \end{Bmatrix} , \tag{11.35}$$

i.e. its mean $\boldsymbol{\mu}_{\boldsymbol{x}}^{(e)}$ and covariance matrix $\boldsymbol{\Gamma}_{\boldsymbol{xx}}^{(e)}$

$$\begin{aligned} \boldsymbol{\mu}_{\boldsymbol{x}}^{(e)} &= E\{\boldsymbol{x}^{(e)}\} , \\ \boldsymbol{\Gamma}_{\boldsymbol{xx}}^{(e)} &= E\{(\boldsymbol{x}^{(e)} - \boldsymbol{\mu}_{\boldsymbol{x}}^{(e)})(\boldsymbol{x}^{(e)} - \boldsymbol{\mu}_{\boldsymbol{x}}^{(e)})^T\} , \end{aligned} \tag{11.36}$$

where (11.35) has again a Karhunen-Loève representation

$$\boldsymbol{x}^{(e)}(t) = \boldsymbol{x}^{(e,0)}(t) + \sum_{j=1}^{3m} \bar{\xi}_j \boldsymbol{x}^{(e,j)}(t) , \tag{11.37}$$

with

$$\begin{aligned} \boldsymbol{u}_L^{(e,j)} &= \boldsymbol{T}^{(e)}\boldsymbol{u}^{(j)} , \\ \dot{\boldsymbol{u}}_L^{(e,j)} &= \boldsymbol{T}^{(e)}\dot{\boldsymbol{u}}^{(j)} , \end{aligned} \tag{11.38}$$

and

$$\boldsymbol{x}^{(e,s,j)}(t) = \begin{Bmatrix} \boldsymbol{u}^{(e,s,j)}(t) \\ \dot{\boldsymbol{u}}^{(e,s,j)}(t) \\ \boldsymbol{q}^{(e,s-1,j)}(t) \end{Bmatrix} . \tag{11.39}$$

For the integration of $\dot{\boldsymbol{q}}^{(j)}$ with respect to time, the simple trapezoidal integration

$${}^{t+\Delta t}\boldsymbol{q}^{(e,j,s)} = {}^{t}\boldsymbol{q}^{(e,j)} + \frac{1}{2}\Delta t[{}^{t}\dot{\boldsymbol{q}}^{(e,j)} + {}^{t+\Delta t}\dot{\boldsymbol{q}}^{(e,j,s)}] \tag{11.40}$$

may often be sufficient. Again, for the beginning of the iteration scheme it is assumed that

$${}^{t+\Delta t}\dot{\boldsymbol{q}}^{(j,0)} := {}^{t}\dot{\boldsymbol{q}}^{(j)} \tag{11.41}$$

holds. Convergence is achieved in case the condition

$$\frac{d}{s} \leq \epsilon \tag{11.42}$$

is satisfied, where ϵ is a suited real number, e.g. $1 \cdot 10^{-6}$, and

$$\begin{aligned} d &= \sum_{i=1}^{n+n_{\boldsymbol{q}}} |\mu_{\boldsymbol{x}[i]}^{(s)} - \mu_{\boldsymbol{z}_i}^{(s-1)}| + |\sigma_{\boldsymbol{x}[i]}^{(s)} - \sigma_{\boldsymbol{x}[i]}^{(s-1)}| , \\ s &= \sum_{i=1}^{n+n_{\boldsymbol{q}}} |\mu_{\boldsymbol{x}[i]}^{(s)} + \mu_{\boldsymbol{z}_i}^{(s-1)}| + \sigma_{\boldsymbol{x}[i]}^{(s)} + \sigma_{\boldsymbol{x}[i]}^{(s-1)} . \end{aligned} \tag{11.43}$$

In (11.43), n and $n_{\boldsymbol{q}}$ denotes the number of degrees of freedom and the number of auxiliary variables, respectively, and $\boldsymbol{x}[i]$ denotes the i-th element of the state vector $\boldsymbol{x}$. To improve the convergence properties of the iterative procedure, e.g. to avoid oscillatory behavior, $\boldsymbol{q}^{(j)}$ might be replaced by a linear combination of the actual and previous iterative solution, i.e.

$$\bar{\boldsymbol{q}}^{(j,s)} = \beta \boldsymbol{q}^{(j,s)} + (1-\beta)\bar{\boldsymbol{q}}^{(j,s-1)} , \tag{11.44}$$

with

$$0 < \beta \leq 1 . \tag{11.45}$$

The value β is set to 1 at the beginning of the iteration and is decreased after two iterations, e.g. to $\beta = 0.5$.

Mode Acceleration Method

According to (6.37), the decoupled equations

$$\begin{aligned} &{}^{t+\Delta t}\ddot{z}_i^{(j,s)} + {}^{t+\Delta t}\dot{z}_i^{(j,s)} 2\zeta_i\omega_i \\ &+ {}^{t+\Delta t}z_i^{(j,s)}\omega_i^2 = \boldsymbol{\phi}_i^T \left({}^{t+\Delta t}\boldsymbol{f}^{(j)} + \boldsymbol{R}^{t+\Delta t}\boldsymbol{q}^{(j,s-1)} \right) , \end{aligned} \tag{11.46}$$

have to be integrated, applying any available and appropriate deterministic step by step integration scheme. For the beginning of the iteration scheme it is assumed that (11.31) holds. Integration of (11.46) yields the associated Karhunen-Loève vectors of the dynamic part of the structural response $\boldsymbol{u}_d$ and $\dot{\boldsymbol{u}}_d$,

$$ {}^{t+\Delta t}\boldsymbol{u}_d^{(j,s)} = \boldsymbol{\Phi}\,{}^{t+\Delta t}\boldsymbol{z}^{(j,s)} , \qquad {}^{t+\Delta t}\dot{\boldsymbol{u}}_d^{(j,s)} = \boldsymbol{\Phi}\,{}^{t+\Delta t}\dot{\boldsymbol{z}}^{(j,s)} . \tag{11.47} $$

In addition, the associated Karhunen-Loève vectors of the static part of the structural response $\boldsymbol{u}_s$ and $\dot{\boldsymbol{u}}_s$ have to be determined according to (6.38)

$$ \begin{aligned} {}^{t+\Delta t}\boldsymbol{u}_s^{(j,s)} &= \hat{\boldsymbol{B}}\,{}^{t+\Delta t}\boldsymbol{a}^{(j)} + \hat{\boldsymbol{R}}\,{}^{t+\Delta t}\boldsymbol{q}^{(j,s-1)} , \\ {}^{t+\Delta t}\dot{\boldsymbol{u}}_s^{(j,s)} &= \hat{\boldsymbol{B}}\,{}^{t+\Delta t}\dot{\boldsymbol{a}}^{(j)} + \hat{\boldsymbol{R}}\,{}^{t+\Delta t}\dot{\boldsymbol{q}}^{(j,s-1)} . \end{aligned} \tag{11.48} $$

Again, for the beginning of the iteration scheme it is assumed that (11.41) holds. Finally, the sums of (11.47) and (11.48) define the Karhunen-Loève vectors of the structural response

$$ \begin{aligned} {}^{t+\Delta t}\boldsymbol{u}^{(j,s)} &= {}^{t+\Delta t}\boldsymbol{u}_d^{(j,s)} + {}^{t+\Delta t}\,\boldsymbol{u}_s^{(j,s)} , \\ {}^{t+\Delta t}\dot{\boldsymbol{u}}^{(j,s)} &= {}^{t+\Delta t}\dot{\boldsymbol{u}}_d^{(j,s)} + {}^{t+\Delta t}\,\dot{\boldsymbol{u}}_s^{(j,s)} . \end{aligned} \tag{11.49} $$

In a next step, the additional variables ${}^{t+\Delta t}\{x_1^{(e,s)}\}_{j=0}^{3m}$ and ${}^{t+\Delta t}\{x_2^{(e,s)}\}_{j=0}^{3m}$ according to (13.29) have to be calculated for every non-linear element, allowing subsequently to determine the linearization coefficients as defined in (13.34)-(13.36) and (13.38)-(13.40). The local element coordinates are related to the reduced global coordinates $(\boldsymbol{z}, \dot{\boldsymbol{z}}, \boldsymbol{q})$ by

$$ \boldsymbol{u}_L^{(e,s)}(t) = \boldsymbol{Q}_1^{(e,s)}\boldsymbol{z}(t) + \boldsymbol{Q}_2^{(e,s)}\boldsymbol{a}(t) + \boldsymbol{Q}_3^{(e,s-1)}\boldsymbol{q}(t) , \tag{11.50} $$

and

$$ \dot{\boldsymbol{u}}_L^{(e,s)}(t) = \boldsymbol{Q}_1^{(e,s)}\dot{\boldsymbol{z}}(t) + \boldsymbol{Q}_2^{(e,s)}\dot{\boldsymbol{a}}(t) + \boldsymbol{Q}_3^{(e,s-1)}\dot{\boldsymbol{q}}(t) . \tag{11.51} $$

Here, the constant transformation matrices read

$$ \boldsymbol{Q}_1^{(e)} = \boldsymbol{T}^{(e)}\boldsymbol{\Phi} , \qquad \boldsymbol{Q}_2^{(e)} = \boldsymbol{T}^{(e)}\hat{\boldsymbol{B}} , \qquad \boldsymbol{Q}_3^{(e)} = \boldsymbol{T}^{(e)}\hat{\boldsymbol{R}} , \tag{11.52} $$

where the element specific transformation vector $\boldsymbol{T}^{(e)}$ has been introduced in (11.17). The matrices $\boldsymbol{Q}_2^{(e)}$ and $\boldsymbol{Q}_3^{(e)}$ stem from the approximation of the response of the higher modes. They are supposed to be small when compared with the effect of the modal quantities. With (11.51) and (11.52) the state vector of the response at element level is available, see (11.35),

$$ \boldsymbol{x}^{(e,s)} = \begin{Bmatrix} \boldsymbol{u}_L^{(e,s)} \\ \dot{\boldsymbol{u}}_L^{(e,s)} \\ \boldsymbol{q}^{(e,s-1)} \end{Bmatrix} . \tag{11.53} $$

thus (11.36)–(11.45) are also valid.

11.3 Non-Zero Mean Response

The described procedure provides also a framework for addressing problems characterized by a non-zero mean response, i.e. $\boldsymbol{u}^{(0)} \neq \boldsymbol{0}$ and $\dot{\boldsymbol{u}}^{(0)} \neq \boldsymbol{0}$. This situation typically arises when there is non-zero mean loading, i.e. $\boldsymbol{f}^{(0)} \neq \boldsymbol{0}$, or when the hysteresis loop shown in Fig. 13.6 shows a non-symmetric behavior. But also for zero mean loading and a symmetric hysteresis loop, the vector $\boldsymbol{q}$ is a non-zero mean vector stochastic process. Since this vector comprises non-decreasing, non-reversible functions of time, a stationary solution does not exist whenever the response reacts non-linearly. In such cases, however, $\mu_{q1}^{(e)} = -\mu_{q2}^{(e)}$ and thus the resulting non-linear restoring force $\boldsymbol{Rq}$ is equal to zero.

While for linear systems the mean and the covariance function, respectively, can be solved independently, they are mutual dependent for non-linear systems. In other words, the zero-th Karhunen-Loève vector of the response affects the variability of the response and thus the higher order Karhunen-Loève vectors, which in turn contribute to the mean response of the system. Similar to Secs. 11.2.2 or 11.2.2, this dependency has to be resolved iteratively, but now two iteration loops are required in order to satisfy equilibrium conditions.

11.3.1 Direct Integration

Starting point for problems with a non-zero mean response is again (6.4). In case of direct integration, (11.30) together with (13.30) and (13.31) has to be solved within the first iteration loop for $j = 0$. Denoting the iteration cycle with r this yields

$$\boldsymbol{M}\,{}^{t+\Delta t}\ddot{\boldsymbol{u}}^{(0,r)} + \boldsymbol{C}\,{}^{t+\Delta t}\dot{\boldsymbol{u}}^{(0,r)} + \boldsymbol{K}\,{}^{t+\Delta t}\boldsymbol{u}^{(0,r)} = {}^{t+\Delta t}\boldsymbol{f}^{(j)} + \boldsymbol{R}\,{}^{t+\Delta t}\boldsymbol{q}^{(0,r-1)} \tag{11.54}$$

and

$$\boldsymbol{q}^{(0,r)} = \int \boldsymbol{g}(\boldsymbol{u}^{(0,r)}, \dot{\boldsymbol{u}}^{(0,r)}, \boldsymbol{q}^{(0,r-1)})dt\,. \tag{11.55}$$

At the beginning of the iteration scheme, i.e. $r = 1$, it is assumed that the condition

$${}^{t+\Delta t}\boldsymbol{q}^{(0,0)} = {}^{t}\boldsymbol{q}^{(0)} \tag{11.56}$$

holds. Once equilibrium is satisfied for (11.54) and (11.55) at iteration cycle $r = R$, the variability of the response can be estimated by computing the higher order Karhunen-Loève vectors according to

$$\boldsymbol{M}\,{}^{t+\Delta t}\ddot{\boldsymbol{u}}^{(j,s)} + \boldsymbol{C}\,{}^{t+\Delta t}\dot{\boldsymbol{u}}^{(j,s)} + \boldsymbol{K}\,{}^{t+\Delta t}\boldsymbol{u}^{(j,s)} = {}^{t+\Delta t}\boldsymbol{f}^{(j)} + \boldsymbol{R}\,{}^{t+\Delta t}\boldsymbol{q}^{(j,s-1)} \qquad j > 0\,, \tag{11.57}$$

where the iteration cycle is now denoted with s. Applying equivalent statistical linearization, the auxiliary variables can be calculated by

$$\boldsymbol{q}^{(0,s)} = \int E\{g(\boldsymbol{u}^{(s)}, \dot{\boldsymbol{u}}^{(s)}, \boldsymbol{q}^{(s-1)})\} \, dt \tag{11.58}$$

and

$$\boldsymbol{q}^{(j,s)} = \int (\boldsymbol{U}\tilde{\boldsymbol{u}}^{(j,s)} + \boldsymbol{V}\dot{\tilde{\boldsymbol{u}}}^{(j,s)} + \boldsymbol{W}\tilde{\boldsymbol{q}}^{(j,s-1)}) dt \, . \tag{11.59}$$

Here, at the beginning of the iteration scheme, i.e. $s = 1$, it is assumed that the conditions

$${}^{t+\Delta t}\boldsymbol{q}^{(0,0)} = {}^{t+\Delta t}\boldsymbol{q}^{(0,R)} \tag{11.60}$$

and

$${}^{t+\Delta t}\boldsymbol{q}^{(j,0)} = {}^{t}\boldsymbol{q}^{(j)} \tag{11.61}$$

hold. As can be seen, the variability of the response contributes to $\boldsymbol{q}^{(0)}$, which requires to update the zero-th Karhunen-Loève-vectors of the response. After equilibrium is satisfied for the second iteration loop at $s = S$, this is done by solving (11.54) and (11.55) iteratively, but now instead of (11.56) the initial condition reads

$${}^{t+\Delta t}\boldsymbol{q}^{(0,0)} = {}^{t+\Delta t}\boldsymbol{q}^{(0,S)} \, . \tag{11.62}$$

For both iteration loops convergence is achieved according to the relations defined in (11.42) and (11.43).

11.3.2 Mode Acceleration Method

If the mode acceleration method is used for the time integration of the problems characterized by a non-zero mean response, everything written in Sec. 11.3.1 is valid and can be applied accordingly.

Part III

Practical Applications

12

Stability Analysis of Cylindrical Shells with Random Imperfections

As an example of imperfection sensitive structures, the axially compressed cylindrical shell is analyzed in the subsequent sections. Despite of its simple geometry, a non-linear static finite element analysis of an isotropic cylindrical shell is from the computational point of view quite demanding. For the determination of the collapse load of the imperfect cylindrical shell a non-linear static finite element analysis using STAGS [104] is carried out. The analysis takes into account non-linearities due to large displacements and large rotations, but is restricted to small strains. The stress-strain relation is assumed to be linear. STAGS is designed for general purpose analysis of shell structures of arbitrary shape and complexity and proves to be most instrumental for involved, non-linear shell problems. It is particularly suited for the use in context with automatized Monte Carlo simulation.

12.1 Stochastic Modeling of Geometric and Boundary Imperfections

The theory of stochastic processes, in particular the theory of random fields, can be applied most advantageously in order to capture the inherent randomness of geometric imperfections of cylindrical shells and other types of shell structures, see Sec. 8.3 and [115, 112].

12.1.1 Estimation of the Covariance Matrix of Geometric Imperfections

With respect to (8.45), geometric imperfections represent a two-dimensional random field

$$g(\boldsymbol{t}); \quad \boldsymbol{t} \in D \subseteq \mathcal{R}^2 , \qquad (12.1)$$

where

$$t = \begin{Bmatrix} \eta \\ \zeta \end{Bmatrix} \tag{12.2}$$

describes a Cartesian coordinate system defined on the surface of the cylindrical shell, see Fig. 12.1. Based on a database of measured imperfections (see

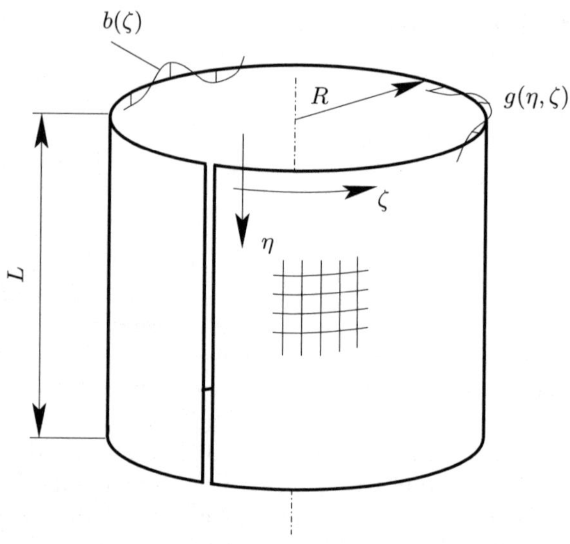

Fig. 12.1. Surface coordinates η, ζ of cylindrical shell. Geometric imperfections $g(\eta, \zeta)$ are defined normal to the shell surface, boundary imperfections $b(\zeta)$ are defined in axial direction. L and R denote the length and the radius of the shell, respectively.

Appendix A), the covariance matrix has been estimated according to (8.13)

$$\boldsymbol{\Gamma}_{gg} = \frac{1}{n-1} \sum_{r=1}^{n} (g_{(r)} - \mu_{g_{(r)}})(g_{(r)} - \mu_{g_{(r)}})^T \} , \tag{12.3}$$

using $n = 7$. Contrary to (8.13), (12.3) represents the best unbiased estimator for the covariance matrix. The structure of the estimated covariance matrix clearly reveals that these measured geometric imperfections represent a non-homogeneous random field, i.e.

$$\Gamma_{gg}((\eta_1, \zeta_1), (\eta_2, \zeta_2)) \neq \Gamma_{gg}(\eta_2 - \eta_1, \zeta_2 - \zeta_1) . \tag{12.4}$$

The covariance function $\Gamma_{gg}(\boldsymbol{t}, \boldsymbol{s})$ between two points (η_1, ζ_1) and (η_2, ζ_2) of the random field thus depends not only on the relative distance between these points but also on the location of these points. The covariance matrix is capable to describe such a shift variant correlation structure without any simplifications or assumptions.

12.1.2 Correlation Structure of Geometric Imperfections

In applied stochastic mechanics the frequently used, so-called correlation length describes in a concise way the correlation structure of a homogeneous random field. It is the purpose of this section – by introducing only for a moment the assumption of homogeneity – to determine the correlation length for the so-called A-shells (see Appendix A).
The correlation length L_c determines the decay of the mutual influence of two different random field locations, e.g. $g(\boldsymbol{t}_1)$ and $g(\boldsymbol{t}_2)$, and is a measure for the number of uncorrelated random variables which are required to describe the random field with a given quality. For an exponential or convex type of covariance function, the correlation length L_c defines the distance where the covariance function $\Gamma_{gg}(\Delta\eta, \Delta\zeta)$ takes on the value $\Gamma_{gg} = \sigma_g^2/e$, where e denotes the Euler number and σ_g^2 the variance of the random field, respectively, and

$$\Delta\eta = \eta_2 - \eta_1, \qquad \Delta\zeta = \zeta_2 - \zeta_1 . \tag{12.5}$$

According to this definition, a correlation length tending towards infinity describes a fully correlated random field, which means that all random variables representing the random field are linearly dependent. A correlation length equal to zero describes an uncorrelated random field, see e.g. [117].

From Figs. A.1(a)–A.4 shown in Appendix A it can be assumed that the variation of the imperfections along the axes η and ζ is independent. This has been substantiated by the data, so it is consistent to imply a fully separable correlation structure of $g(\boldsymbol{t})$. Under the assumption of zero mean and homogeneity the covariance function can then be expressed as, see e.g. [153],

$$\Gamma_{gg}\left(\Delta\eta, \Delta y\right) = \sigma_g^2 \varrho_\eta\left(\Delta\eta\right) \varrho_\zeta\left(\Delta\zeta\right) , \tag{12.6}$$

where ϱ_η, ϱ_ζ are the correlation coefficient functions along the axes η and ζ. This allows to split the correlation structure of the geometric imperfections into two one dimensional problems. In Figs. 12.2 and 12.3 the different correlation coefficient functions for the axial and circumferential direction are shown. The shape of the correlation coefficient function ϱ_ζ can be explained as follows. Its symmetry is due to the fact, that the maximum lag $\Delta\zeta$ is half of the circumference, while the extremes are due to the low frequencies ruling the imperfections in the circumferential direction. Because of the fully separable correlation structure of the imperfections, the correlation length can now be defined for the axial and the circumferential direction individually. According to the definition of the correlation length mentioned before, for the measured geometric imperfections the values

$$L_{c_\eta} \approx 0.6L , \qquad L_{c_\zeta} \approx 0.06(2R\pi) , \tag{12.7}$$

have been obtained, where L_{c_η} and L_{c_ζ} denote the correlation lengths in axial and circumferential direction, respectively.

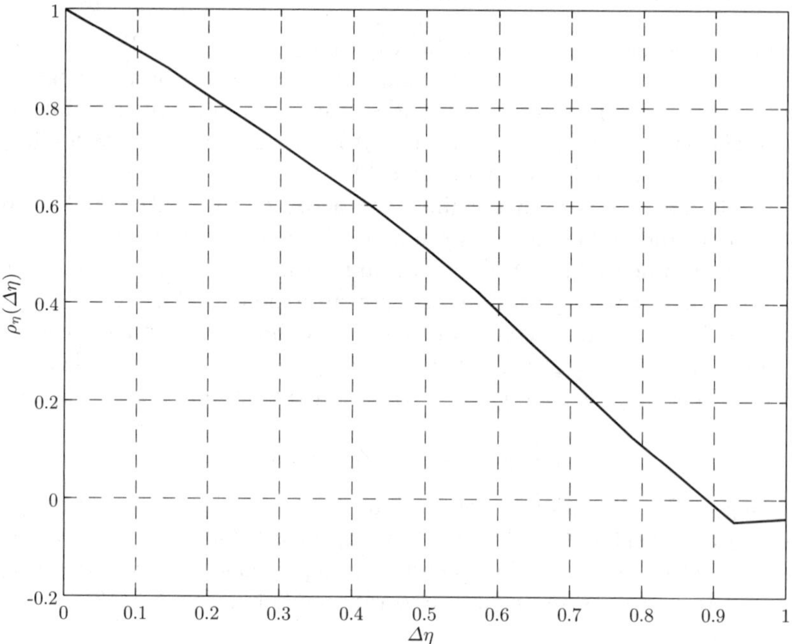

Fig. 12.2. Correlation coefficient function ($\varrho_\eta(\Delta\eta)$, $\Delta\eta = [0, L]$) in axial direction of the A-shells under the assumption of homogeneity and a fully separated correlation structure.

The two dimensional covariance function $\Gamma_{gg}(\Delta\eta, \Delta\zeta)$ of the imperfections according to (12.6) is shown in Fig. 12.4. A separable correlation structure assures a so called quadrant symmetry, i.e. the covariance function defined for positive lags $\Delta\eta$ and $\Delta\zeta$ fully represents the correlation structure of the imperfections $g(\boldsymbol{t})$ [153].

12.1.3 Estimation of the Covariance Function of Boundary Imperfections

Boundary imperfections of cylindrical shells will be considered in the following and are defined, according to Fig. 12.1, by

$$b(\boldsymbol{t}); \quad \boldsymbol{t} \in D \subseteq \mathcal{R}^1 , \tag{12.8}$$

where

$$\boldsymbol{t} = \{\zeta\} . \tag{12.9}$$

Second moment characteristics of boundary imperfections cylindrical shells have been estimated from results of a flatness survey (see Fig A.5 and Fig. 4

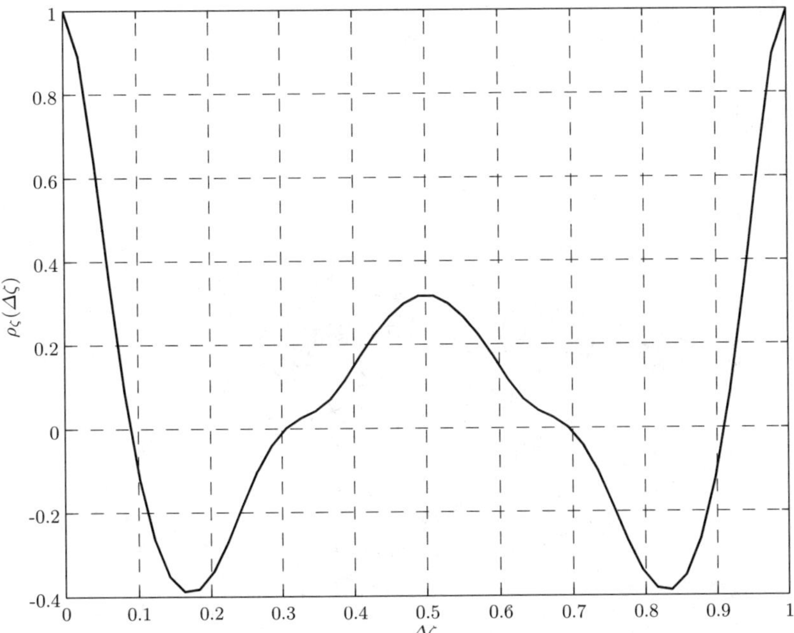

Fig. 12.3. Correlation coefficient function ($\varrho_\zeta(\Delta\zeta)$, $\Delta\zeta = [0, 2R\pi]$) in circumferential direction of the A-shells under the assumption of homogeneity and a fully separated correlation structure.

in [4]) of one of the end-rings used in a test setup for buckling experiments of anisotropic shells. Since this is the only sample that has been available, ergodicity is assumed, i.e. the covariance function $\Gamma_{bb}(\zeta_2 - \zeta_1) = \Gamma_{bb}(\Delta\zeta)$ shown in Fig. 12.5, has been calculated by spatial averaging, see also [113]. Its symmetry is due to the fact, that the maximum lag $\Delta\zeta$ is half of the circumference.

12.2 Modeling of Imperfect Shell Structures by Finite Elements

Three different cylindrical shells are analyzed next, see Table 12.1 and 12.2, respectively: A very thin cylindrical shell in presence of geometric and/or boundary imperfections (shell #1, see Fig. 12.6 and [114]), and two shells with a cutout in presence of geometric imperfections (shell #2 and #3, see Fig. 12.7 - 12.8 and [113, 116]).

All finite element models are established by an assumed natural strain 9-node quadrilateral shell element [88] with 5 degrees of freedom per node. This

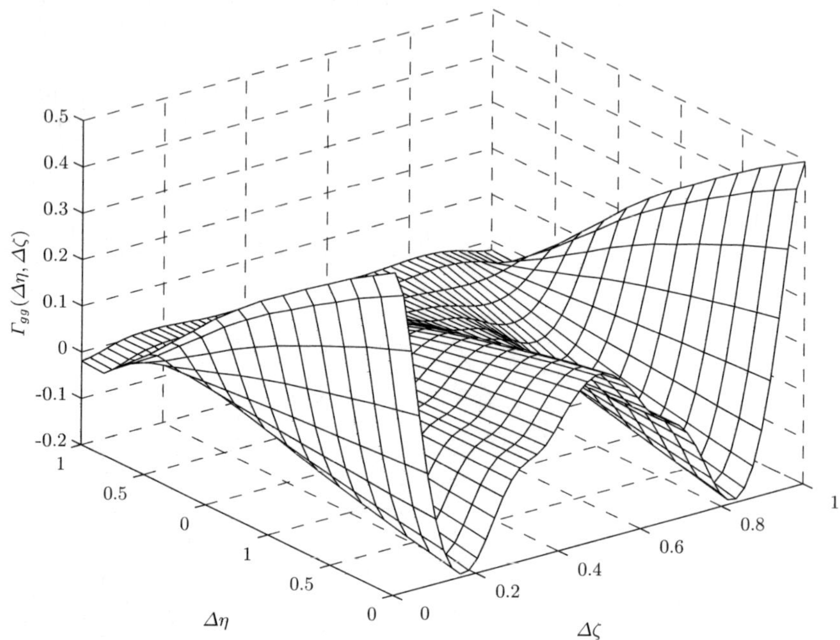

Fig. 12.4. Two dimensional covariance function $\Gamma_{gg}(\Delta\eta, \Delta\zeta)$ of the A-shells under the assumption of homogeneity and a fully separated correlation structure.

Shell	R [m]	t [m]	L [m]	E [N/m^2]	ν [-]
#1	0.1016	0.00011597	0.2023	$1.0441{\cdot}10^{11}$	0.3
#2	0.1016	0.00115970	0.2023	$1.0441{\cdot}10^{11}$	0.3
#3	0.1016	0.00115970	0.2023	$1.0441{\cdot}10^{11}$	0.3

Table 12.1. Dimensions and material properties of finite element models. Radius R, thickness t, length L, Young's modulus E and Poisson's ratio ν.

element is referred to as '480' element in [104], in particular 9 Gauss integration points has been used. Contrary to other shell elements used for modeling very thin shell structures, this element does not show so called "hourglass-modes" and performs very well with regard to an accelerated convergence [103].

On both edges '1' and '3' of the shell shown in Fig. 12.9, the classical simply-supported (SS-3) boundary condition is employed, i.e. $v = w = 0$ and $u \neq 0$,

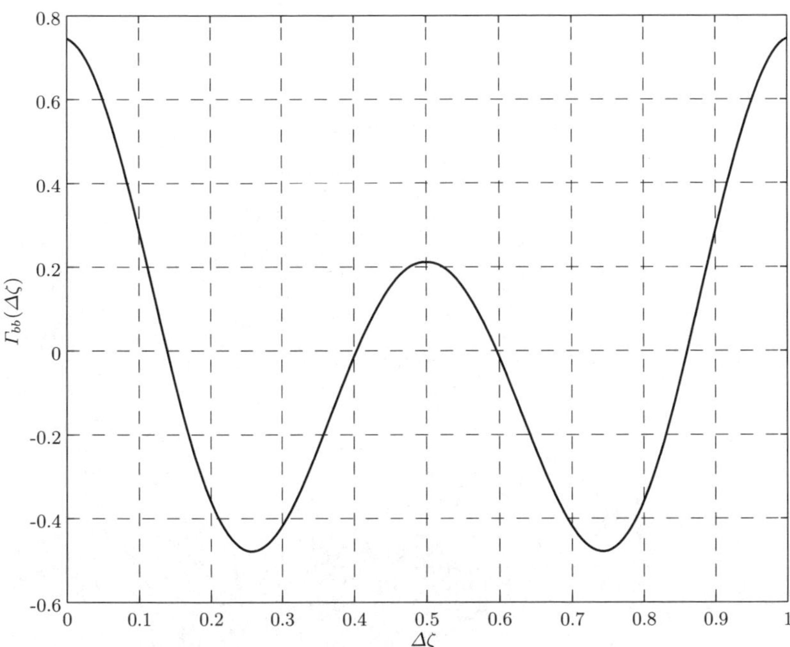

Fig. 12.5. Covariance function $\Gamma_{bb}(\Delta\zeta)$ of boundary imperfections under the assumption of homogeneity.

Shell	H [m]	H/W [-]	$\lambda_{\text{lin}}^{\text{bif}}$	$\lambda_{\text{nl}}^{\text{bif}}$	$\lambda_{\text{nl}}^{\text{lim}}$
#1	-	-	1.0	0.87382	-
#2	0.0154	2.564	0.24390	0.48736	0.65619
#3	0.0456	2.564	0.45387	0.53338	0.56691

Table 12.2. Cutout height H, aspect ratio H/W, bifurcation load obtained by linear analysis $\lambda_{\text{lin}}^{\text{bif}}$, bifurcation load obtained by non-linear analysis $\lambda_{\text{nl}}^{\text{bif}}$, limit load obtained by non-linear analysis $\lambda_{\text{nl}}^{\text{lim}}$.

$ru \neq 0$. In order to ensure an uniform axial displacement of edges '1' and '3', additional constraints have been imposed using the Lagrange multiplier technique.

The use of a Fourier series for the representation of deterministic or random imperfections has been justified in the past mainly because of two reasons. The first one is that in an analytical buckling analysis of cylindrical shells imperfections can only be incorporated if they are similar to critical buckling modes. The second one is that unfortunately only few measurements of imperfections do exist, hence the best choice is again to model imperfections similar to some

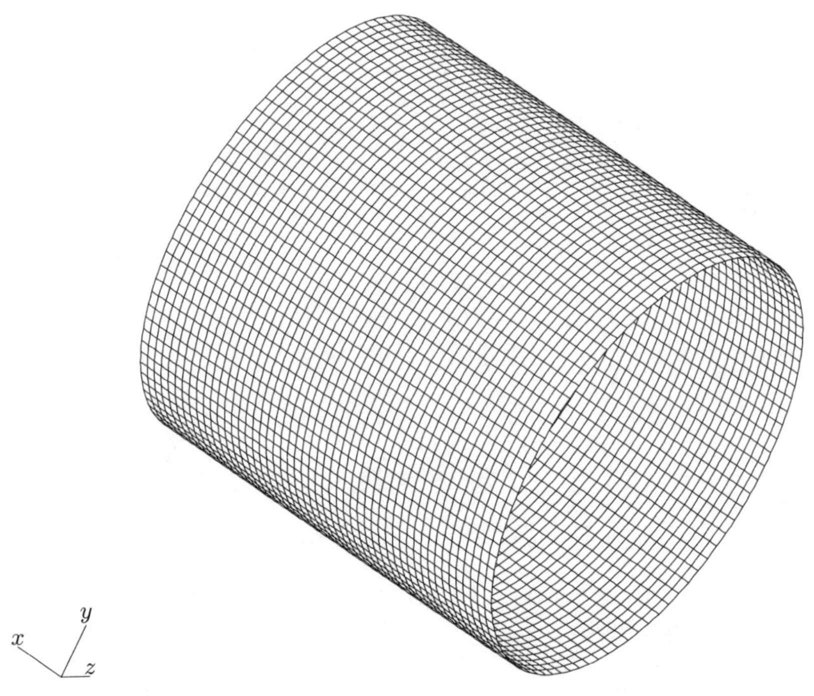

Fig. 12.6. Finite element model of shell #1: mesh 101 × 193.

critical buckling modes (worst case analysis). For an uncertainty analysis of cylindrical shells, it is thus quite obvious that random geometric imperfections have been represented in the past by using random trigonometric polynomials, in particular half-wave cosine or half-wave sine representations. Moreover this approach allows to incorporate only those modes of imperfections, which are expected to have a crucial influence on the buckling behavior of the shell. However, nowadays highly sophisticated numerical methods, such as the finite element method are available for the determination of the critical load, which provide extensive modeling capabilities and allow the incorporation of general geometric imperfections in terms of nodal displacements. Thus there is no need anymore for the transformation of measured imperfections from the space domain into the spatial frequency domain.

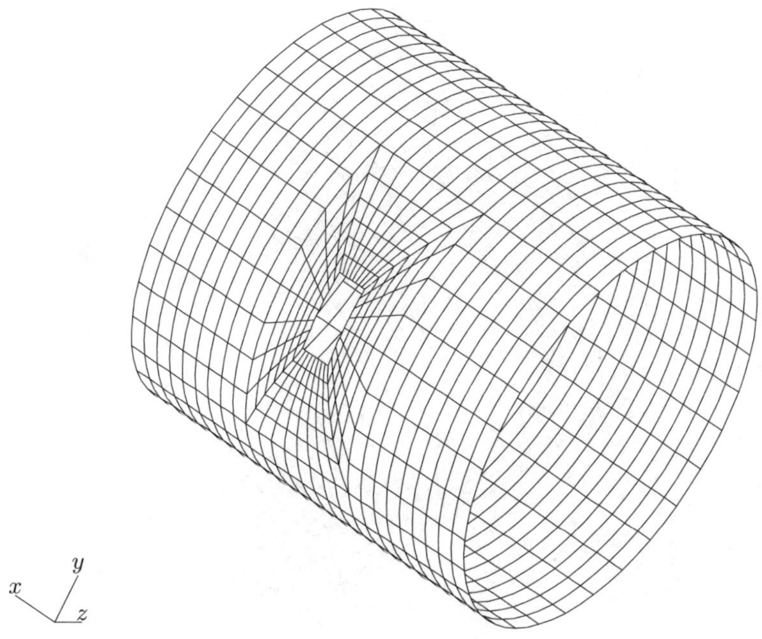

Fig. 12.7. Finite element model of shell #2: mesh 53×65.

Dealing with random imperfections, the Karhunen-Loève expansion is perfectly suited for a stability analysis in context with the finite element method, because no simplifications of the shape of imperfections are introduced. A minor disadvantage of such an approach is the necessity to model the complete shell due to missing symmetry planes, which of course increases the computational time needed.
Geometric imperfections are defined as nodal displacements in the w-direction representing perturbations of the idealized shell geometry, whereby higher order strains, as e.g. used in [108], are not considered in the following.
Boundary imperfections are modeled as a full contact problem using a two step approach. The first step introduces boundary imperfections a prescribed displacements in the u-direction on both edges of the shell. This step introduces rather localized stress fields close to the edges of the shell. In the second step, the external nodal forces as a result of the first step are applied on both edges of the shell together with the axial loading. The resulting effect of this two step approach is a non-uniformly axial loading, governed by the shape of the boundary imperfections.

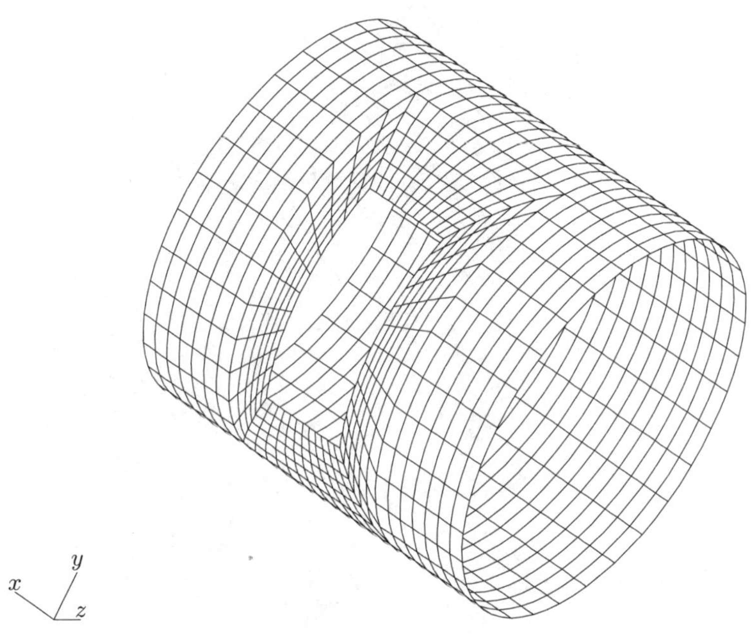

Fig. 12.8. Finite element model of shell #3: mesh 53 × 65.

12.3 Buckling Load of Perfect Shell

12.3.1 Cylindrial Shell without Cutout

In the following, all load levels are normalized by the classical buckling load of an axially compressed isotropic shell with SS-3 boundary condition on both edges after [74] and [150],

$$N_{cl} = E\,(3\,(1-\nu^2))^{-0.5}\,t^2\,R^{-1}\,. \tag{12.10}$$

Isotropic shells under axial compression do have the peculiarity that the non-uniformity and non-linearity of the pre-buckling state near the edges reduces the predicted critical load considerably, moreover the asymmetric bifurcation buckling load is very close to the axi-symmetric buckling load (see e.g. [25]). Keeping this in mind, the two lowest eigenvalues obtained by a non-linear bifurcation analysis carried out by STAGS together with their corresponding buckling modes have been compared with semi-analytical results available, see e.g. [13].

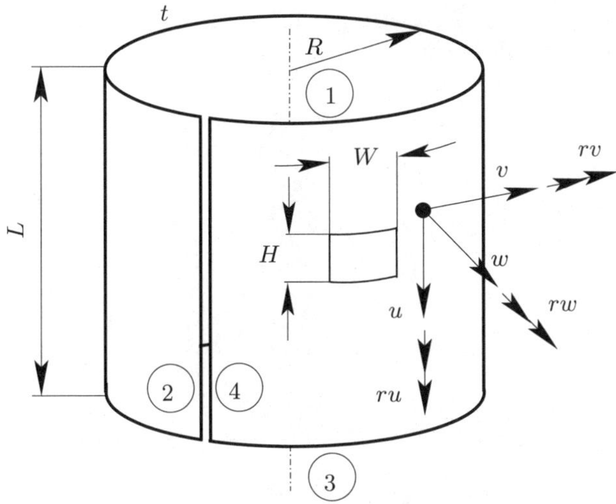

Fig. 12.9. Cylindrical shell: radius R, thickness t, length L, cutout height H, cutout width W, see Table 12.1.

For this purpose the batch tool ANILISA [6] as part of SDAS [62], proves to be most instrumental. ANILISA is based on Donnell type equations where first the axi-symmetric non-linear pre-buckling path of the perfect shell is calculated. After a range of circumferential wave numbers has been selected, load levels at which bifurcation occurs are calculated. This is achieved by solving numerically the resulting eigenvalue problem, whereby the user selected boundary conditions, in this case SS-3, are satisfied rigorously.

Dimensions and material properties of the finite element model are given in Table 12.1 and 12.2, respectively, in particular shell #1 is obtained by averaging the corresponding parameters available in [5] and included in Table A.1. The two lowest bifurcation loads as obtained by ANILISA for such a shell are

$$\lambda_{\mathrm{nl,sym}}^{\mathrm{bif}} = 0.844343 \quad (n = 24), \quad \lambda_{\mathrm{nl,asym}}^{\mathrm{bif}} = 0.844343 \quad (n = 24)\,, \qquad (12.11)$$

where n denotes the number of circumferential full-waves. The lowest buckling loads occur at a relatively high wave number ($n = 24$) in the circumferential direction and do involve very short wave patterns in the axial direction, close to the edges of the shell. In Fig. 12.10(a) the characteristic pre-buckling deformation of the perfect shell as obtained by ANILISA is shown. The axi-symmetric buckling mode with respect to $L/2$ is shown in Fig. 12.10(b). It is quite obvious, that the finite element mesh has to be rather fine in order to be able to predict the correct buckling behavior of the perfect shell. In this regard, at least two elements or five grid points should represent a half wave, yielding 193 grid points for the circumferential direction.

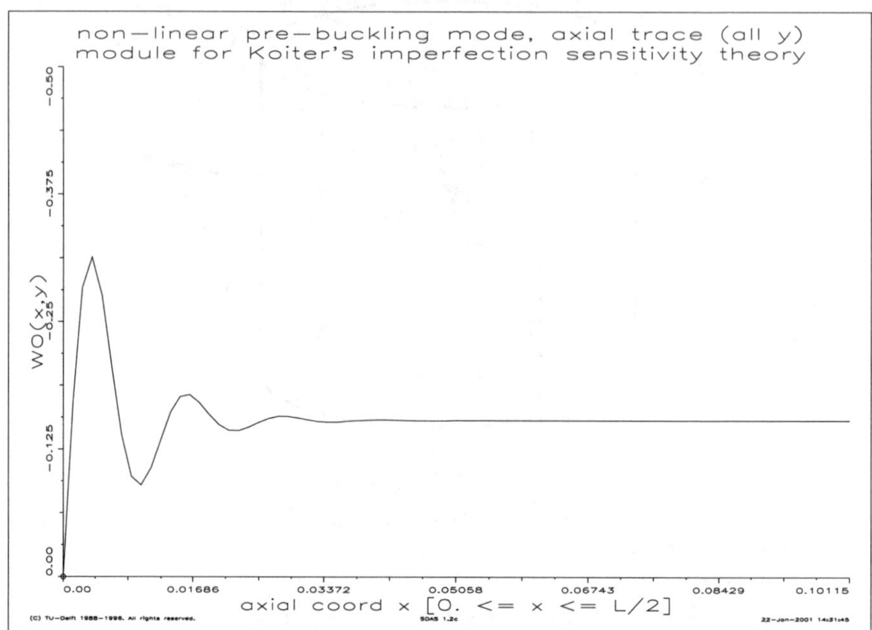

(a) Axial trace of pre-buckling deformation.

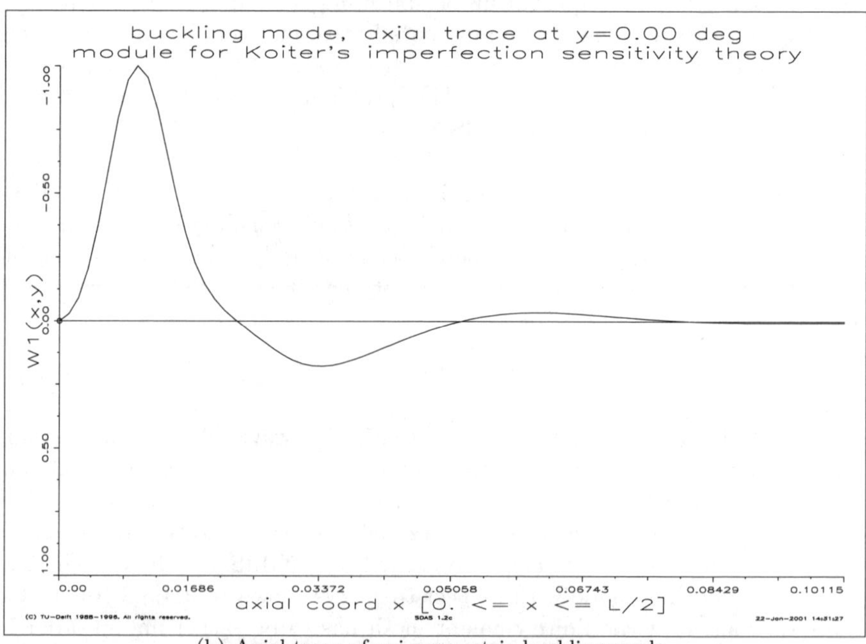

(b) Axial trace of axi-symmetric buckling mode.

Fig. 12.10. Pre-buckling deformation and buckling mode for perfect cylindrical shell by SDAS.

A convergence study, see Fig. 12.11, where the number of nodes in both circumferential and axial direction is varied, revealed that the mesh 101×193 predicts the correct behavior of the perfect shell with a rather small discretization error (~3%), i.e.

$$\lambda^{\text{bif}}_{\text{nl,sym}} = 0.873822 \quad (n = 24), \qquad \lambda^{\text{bif}}_{\text{nl,asym}} = 0.873825 \quad (n = 24)\,. \quad (12.12)$$

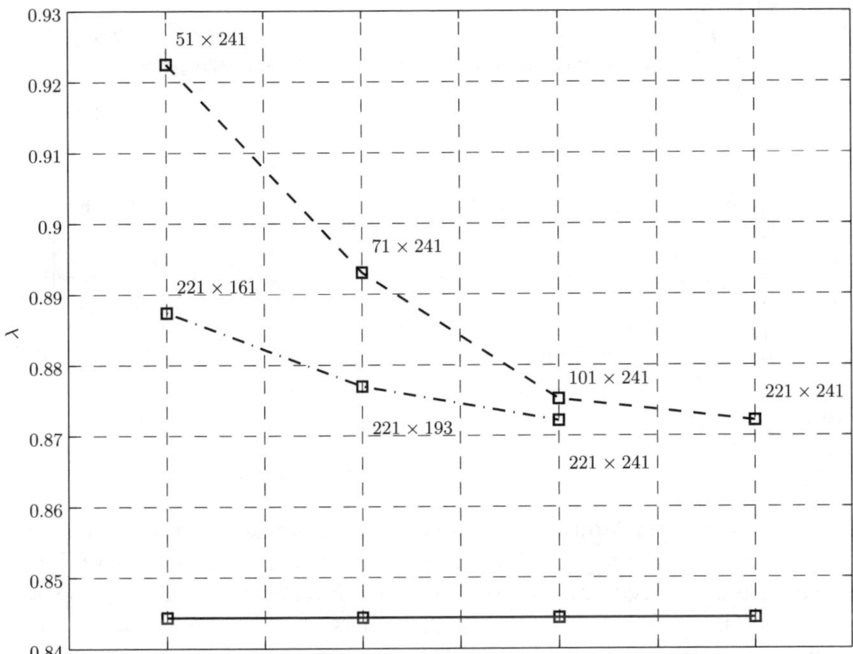

Fig. 12.11. Two dimensional convergence study using STAGS of the perfect shell in axial direction (dashed) and circumferential direction (dash-dot). Labels denote finite element mesh (axial grid points × circumferential grid points). Semi-analytical solution by SDAS (solid).

The pre-buckling deformation as obtained by STAGS is shown in Fig. 12.12, the lowest axi-symmetric buckling mode is plotted in Fig. 12.13(a) and 12.13(b). As can be seen, these plots do agree very well with the corresponding ANILISA plots. It seems to be justified to assume that the response of the imperfect shell is also computed accurately by using this mesh. It should be noted that the mesh 101 × 103 yields a finite element model of approximately 100,000 degrees of freedom. Even nowadays, this is – particularly in

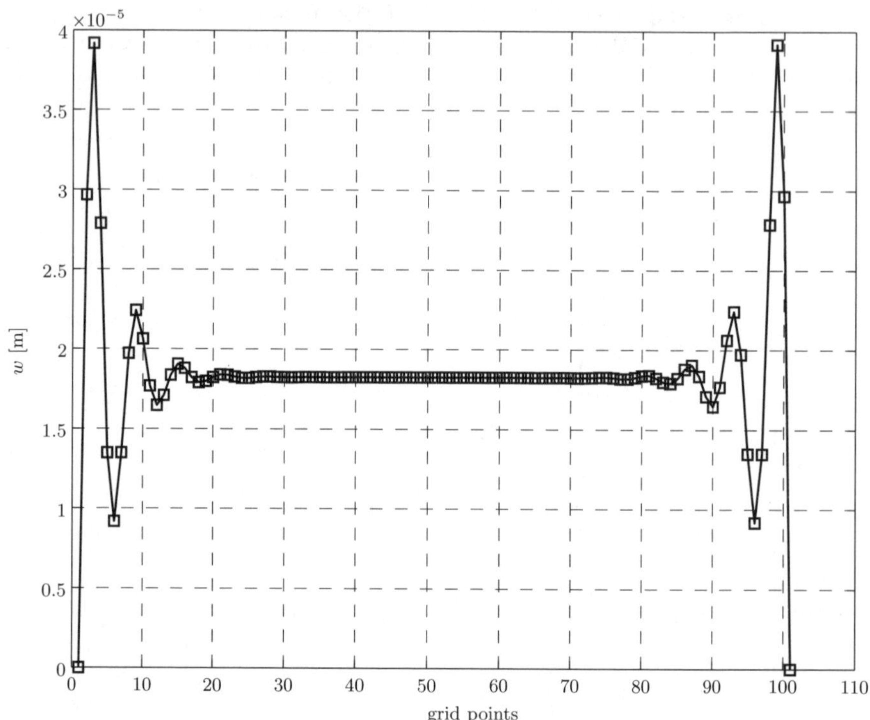

Fig. 12.12. Axial trace of pre-buckling deformation of the perfect shell by STAGS.

context with direct Monte Carlo simulation – computationally quite demanding. Therefore, in order to reduce the number of degrees of freedom of the finite element model, also non-uniformly meshes in axial direction, as proposed in [24], have been tested. Contrary to [24], the size of an element has been continuously increased, with very small elements near the edges of the shell and a rather coarse mesh around $L/2$. However, while for the perfect shell the computational effort can indeed be reduced considerably, for imperfect shells this approach leads to rather inaccurate results. One of the reasons for this might be, that for imperfect shells an isolated buckle may occur, governed by the shape of geometric imperfections, not only close to the edges of the shell but also in the region where the mesh is rather coarse. A "detrimental" buckling mode can obviously not be represented with such a coarse mesh.

12.3.2 Cylindrial Shell with Cutout

Dimensions of shells #2 and #3 are given in Table 12.1 and 12.2, respectively. Again, all load levels are normalized by the classical buckling load of an axially compressed isotropic shell with SS-3 boundary condition on both ends, see (12.10).

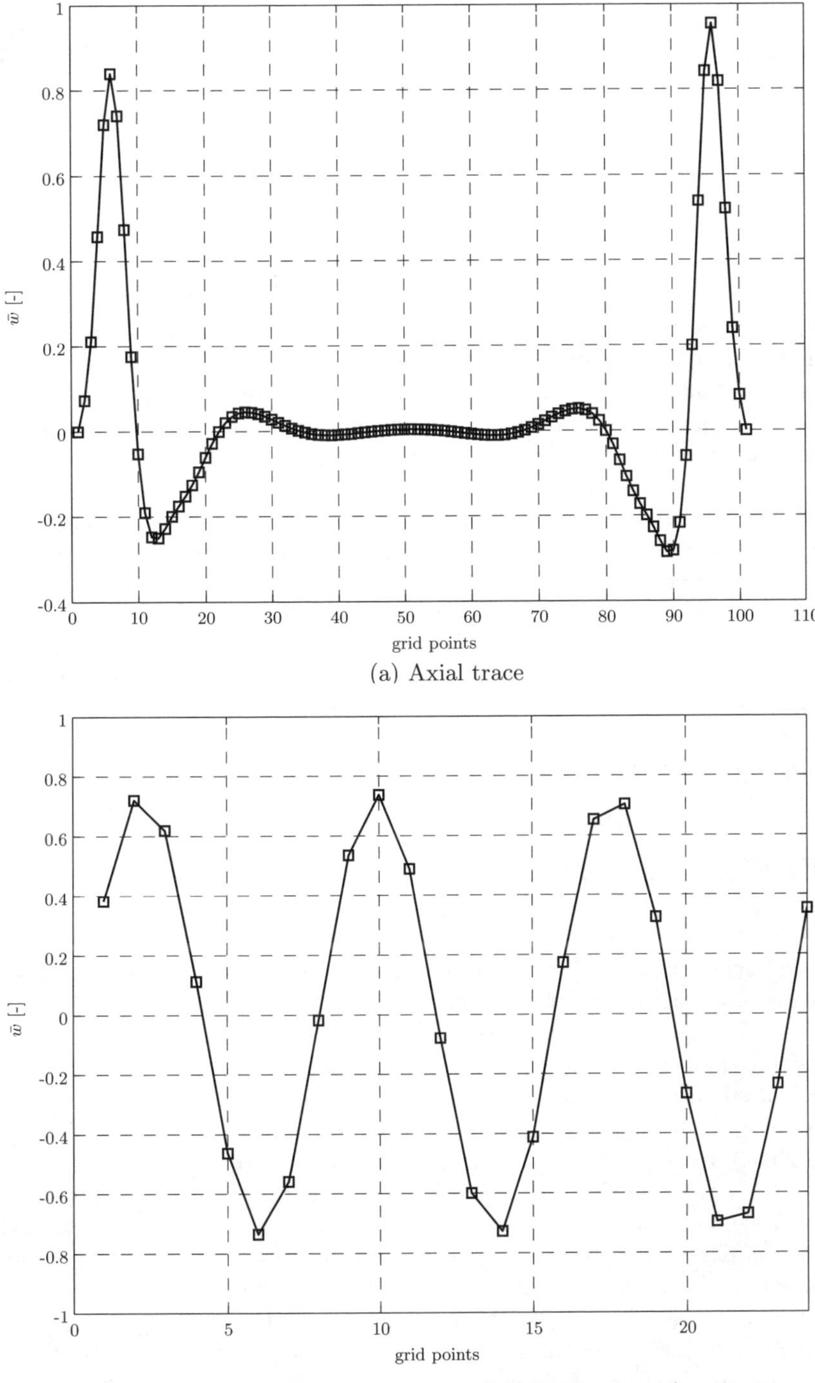

(a) Axial trace

(b) Circumferential trace (45° shell-segment).

Fig. 12.13. Axi-symmetric buckling mode of the perfect shell by STAGS.

First, a two-dimensional convergence study has been carried out for the shell with no cutout, showing that the mesh 53×65 ($\sim$ 25,000 degrees of freedom) predicts the behavior of the perfect shell acceptably well. The discretization error in this case is about 2.5% when compared with semi-analytical results predicted by ANILISA. The lowest buckling load $\lambda_{\mathrm{nl}}^{\mathrm{bif}}$ as obtained by a non-linear analysis carried out by STAGS with the aforementioned discretization is

$$\lambda_{\mathrm{nl}}^{\mathrm{bif}} = 0.8658071 \quad (n = 8) , \tag{12.13}$$

where n denotes the number of circumferential full waves. The effect of a cutout on the buckling load of a cylindrical shell under axial compression is at first sight quite obvious: the cutout weakens the structure. However, due to geometry change in the cylinder the strain energy from the rapidly deforming regions is transferred to other regions which are capable to withstand more axial compression before becoming unstable, see e.g. [25]. This is the reason why in this case a linear bifurcation analysis ($\lambda_{\mathrm{lin}}^{\mathrm{bif}}$) underestimates failure ($\lambda_{\mathrm{nl}}^{\mathrm{bif}}$) considerably, see Table 12.2.
For a realistic prediction of the buckling behavior and to minimize the discretization error, the finite element mesh has been refined near the edges of the cutout. In order to ensure that the mesh transition close to the cutouts does not introduce artificial imperfections, the results obtained using this mesh have been verified with a 100,000 degrees of freedom finite element model using the standard uniform meshing available in STAGS. According to Table 12.2, for both cutouts there is a single bifurcation point $\lambda_{\mathrm{nl}}^{\mathrm{bif}}$ along the equilibrium path before the limit load $\lambda_{\mathrm{nl}}^{\mathrm{lim}}$ is reached.

12.4 Buckling Load of Imperfect Shells

Due to the highly non-linear buckling behavior of imperfect, axially compressed cylindrical shells, direct Monte Carlo simulation seems to be the only possible approach in order to obtain reliable statistical characteristics of the limit load. This is particularly true when several sources of imperfections are considered, since their interaction is rather involved. Thus it is clear, that accelerated Monte Carlo simulation techniques such as variance reduction methods, which themselves require information on the region where failure is most likely to occur, can not be applied in an effective way.
As already mentioned above, a non-linear static finite element analysis with up to 100,000 degrees of freedom is a quite demanding task. In order to determine reliable second order characteristics of the limit load by means of direct Monte Carlo simulation, several finite element analyses are necessary, the resulting computational efforts are enormous. It is clear that the feasibility of such an undertaking depends on a number of factors, such as an optimized solution control for the non-linear analysis, an efficient simulation procedure for generating realizations of geometric imperfections, an appropriate computer

hardware and a fully automatized simulation procedure allowing to connect the different software packages needed without user interaction.
In this context, besides STAGS, also the codes STAR [22], MATLAB [76] and PERL [90], respectively, have been used. The generation of realization of geometric and boundary imperfections has been written in MATLAB and performed on a PC with 450 MHz One MATLAB run took a few seconds only and hence can be neglected when compared with the overall processing time. The computationally demanding non-linear static analysis by STAGS has been performed on a SUN Solaris ULTRA 2/400. Using a Network Queuing System, it was possible to vary the number of jobs running simultaneously depending on the time of the day.
The non-linear static analysis was terminated when passing the limit load level. In a subsequent run a bifurcation analysis based on that non-linear pre-stress state has been carried out. Therefore, the input files of the bifurcation analysis had to be modified in accordance to the results of the foregoing static analysis without user interaction. For post-processing purposes the post-processor STAPL as part of the STAGS standard distribution has been used as well as MATLAB. Since the complete simulation procedure has been distributed over geographically different locations, the PERL has been used for carrying out the necessary network administration tasks, e.g. for data transfer and synchronization of the codes STAGS, STAR and MATLAB.

12.4.1 Cylindrical Shell with Random Geometric Imperfections

Geometric imperfections have been measured with meshes in the range of 15×49 and 31×49, see Table A.1, which are considerably coarser than the mesh required for being able to represent the critical buckling modes. Therefore, the simulated imperfections have been mapped by bi-cubic interpolation to the actual finite element mesh.
In presence of imperfections, the bifurcation point of a perfect cylindrical shell may be replaced by the limit point, i.e. failure occurs as a snap-through when the load level reaches the limit load. This limit load, also often referred as collapse load, has been identified in a non-linear static analysis as the maximum of the load-deflection curve. The starting load factor and starting load factor increment for all non-linear static analyses carried out was $0.05N_{Cl}$ and $0.1N_{Cl}$, respectively, see (12.10). After the first two load steps the solution control was switched from load control to arc-length control. For the solution of the non-linear equation system the modified Newton-Raphson scheme has been applied.
As mentioned in [104], it is possible to gain some more insight about the size of the margin of safety against collapse through the execution of an analysis of bifurcation from the non-linear stress state corresponding to the limit load, in particular if convergence difficulties occur. The so obtained eigenvalue represents according to (5.19) an estimate of the location of the limit point, the

corresponding eigenmode represents physically an estimate of the increment of the deformations at the limit load.

Vicarious for all non-linear static analyses carried out, the deformation of shell #1 with geometric imperfections according to shell A-8 [5] is shown. While Fig. 12.14 represents the deformation of the shell in the pre-buckling state, Fig. 12.15 shows the deformation pattern at the limit load. Fig. 12.16 shows the first eigenmode belonging to the lowest eigenvalue obtained by a bifurcation analysis based on the non-linear stress state at the limit load.

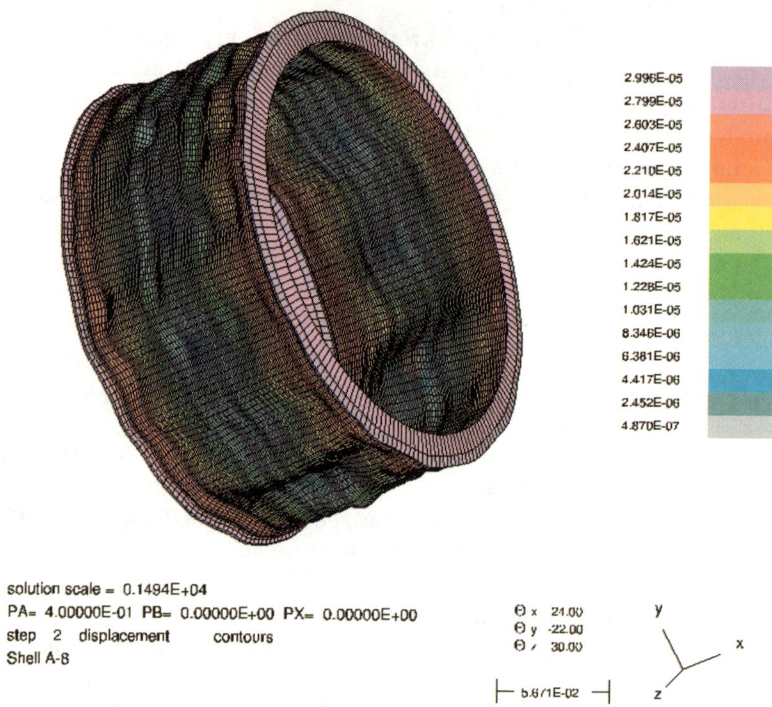

Fig. 12.14. Deformation (prebuckling) for shell A-8 by STAGS.

Monte Carlo simulation has been carried out with 250 realizations of geometric imperfections generated by the Karhunen-Loève expansion, the numerically predicted limit loads by a non-linear analysis as well as estimates of the limit load obtained by a buckling analysis are shown in Fig. 12.17.

One analysis, i.e. a non-linear and a bifurcation analysis, respectively, took about 7,000 CPU seconds on a a SUN Solaris ULTRA 2/400. The overall processing time (at least two analyses run simultaneously) for these 250 non-linear and bifurcation analyses, respectively, was about two weeks. As a final result the histogram of the calculated limit loads is presented in Fig. 12.18(a).

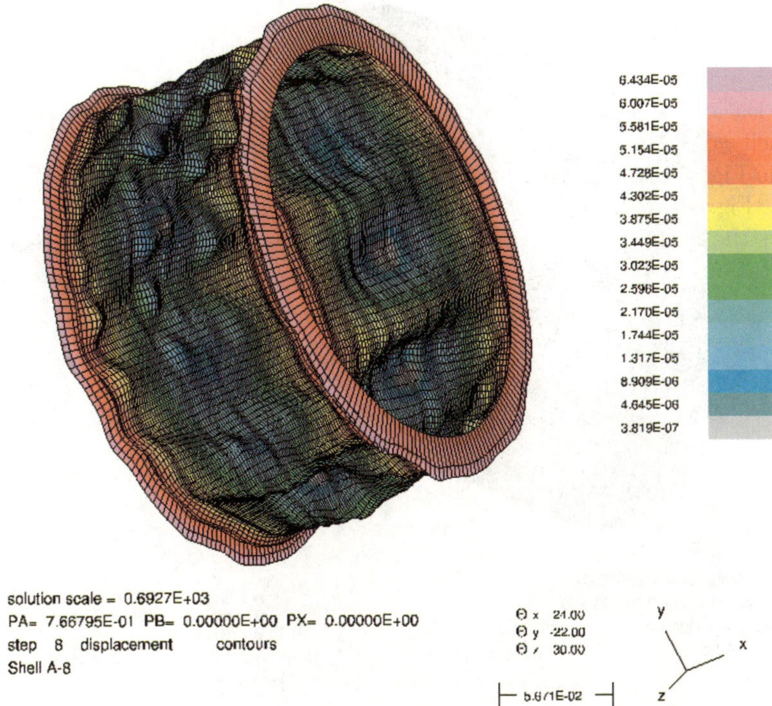

Fig. 12.15. Deformation at the limit point for shell A-8 by STAGS.

By comparing Fig. 12.18(a) with Fig. 12.19(b) it can be observed, that the experimentally determined scatter in the limit load can be predicted numerically. The coefficient of variation of the limit load calculated by STAGS is $V_{\mathrm{STAGS}} = 0.0820$, whereby for the A-shells the corresponding value is $V_{\mathrm{exp}} = 0.0867$. The calculated mean value $\mu_{\mathrm{STAGS}} = 0.7793$ of the limit load is considerably higher when compared with the experimentally determined counterpart $\mu_{\mathrm{exp}} = 0.6430$. This can be explained by the fact, that the test specimen of course do have in addition to geometric imperfections also thickness imperfections, varying material properties (Young's modulus), non-perfect boundary conditions and misalignments in the loading. Geometric and boundary imperfections, respectively, are considered in Sec. 12.4.2, nevertheless, as shown, geometric imperfections do have a strong effect on the limit load of cylindrical shells.

An interesting insight into the stability behavior of cylindrical shells from the reliability point of view provides Fig. 12.20. Here, the ξ's of the Karhunen-Loève expansion (representing the standard normal space) for realizations of geometric imperfections, which have a most detrimental effect on the limit load ($\lambda_{\mathrm{nl}}^{\mathrm{lim}} < 0.6$) are shown. The ξ's belonging to a particular realization are

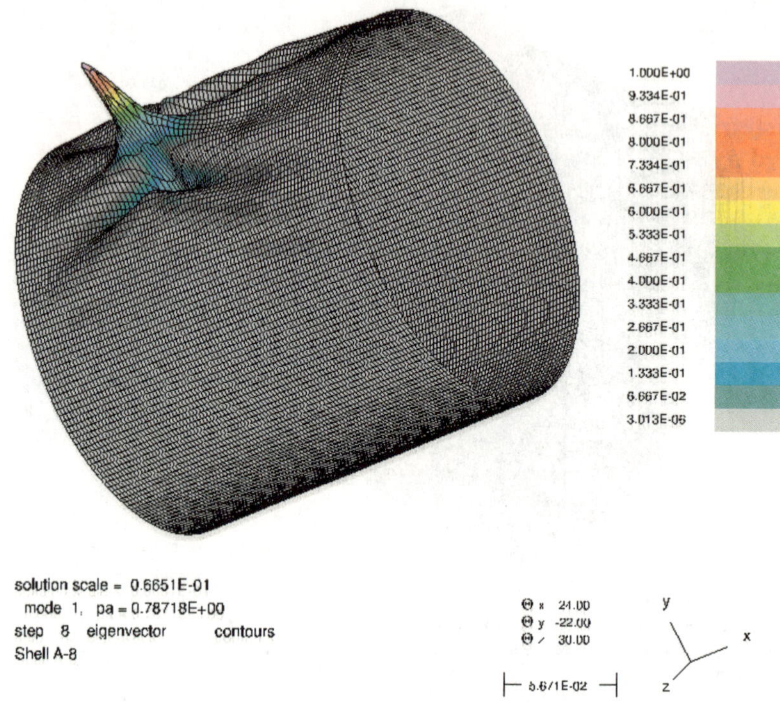

Fig. 12.16. First eigenmode for shell A-8 by STAGS at the limit load, representing the growth of the deformations at the limit point.

connected by a dotted line. As can be seen, it appears that there is no tendency regarding a certain constellation of the ξ's which lowers the limit load significantly. This observation shows again that direct Monte Carlo simulation seems to the only possible method of solution for such highly non-linear systems.

12.4.2 Cylindrical Shell with Random Geometric and Boundary Imperfections

In order to verify the two step approach as described in Sec. 12.2, a type of "worst case" study is carried out, i.e. one triggers the critical buckling load with 24 full waves of the perfect shell with an appropriate shape of boundary imperfections on the shell boundaries '1' and '3', see Fig. 12.9. That is,

$$b_1 = \alpha\, t \cos 24\zeta, \qquad b_3 = -\alpha\, t \cos 24\zeta. \tag{12.14}$$

The presence of harmonically varying boundary imperfections given by (12.14) (symmetric with respect to $L/2$) and a magnitude of one-tenth of the wall thickness, i.e. $\alpha = 0.1$, reduces the buckling of the perfect shell about 71%,

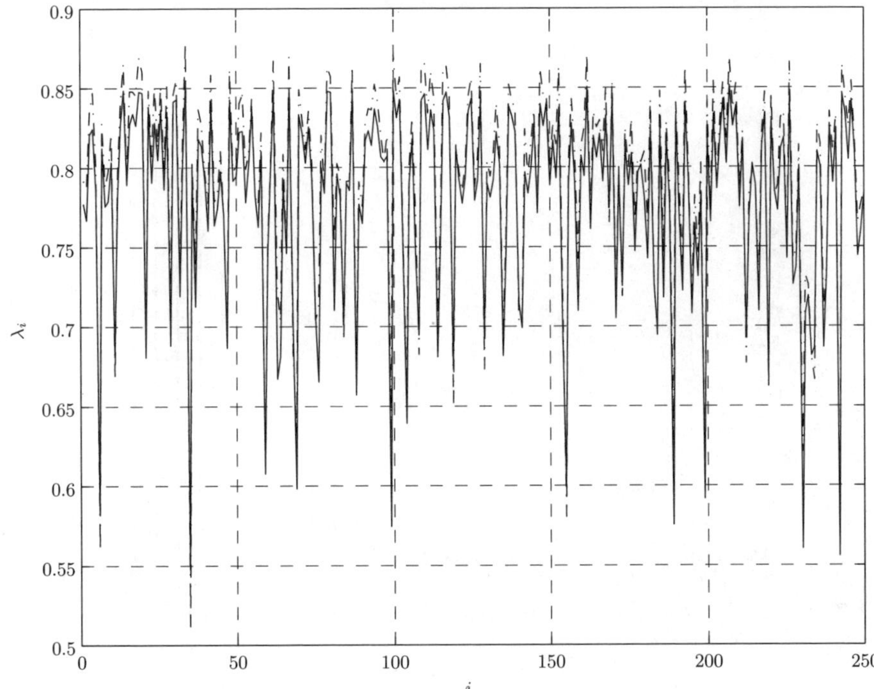

Fig. 12.17. Monte Carlo simulations of the limit load for shell #1: non-linear analysis (solid), bifurcation analysis (dash-dot), i denotes the number of simulation.

which is even more than reported in [4] for anisotropic shells.
When comparing the shape of a boundary imperfection as shown in Fig. A.5 with the idealized shape given in (12.14), two remarks are in order. Firstly, the simulated boundary imperfections are governed by rather low frequencies and secondly, the magnitudes of the simulated boundary imperfections are considerably larger. The first statement implies that the simulated boundary imperfections are unlikely to trigger the critical buckling mode with n=24 full waves, whereby the larger magnitude of the simulated imperfections would reduce the limit load in an unrealistic way, keeping in mind, that the lowest measured limit load observed for the so-called A-shells [5] was about λ=0.565. This contradiction could be explained in the way boundary imperfections are modeled in the non-linear analysis as compared to the experimental setup. As mentioned in Sec. 12.2, the 2-step approach simulates a full contact problem between the end-rings and the shell boundaries. In the experimental setup this assumption has been met only partially. Here, the effect of boundary imperfections have been attempted to be compensated by filling in both edges of the shell with a special material for the purpose to ensure uniform loading conditions over the entire circumference. However, the Young's modulus of

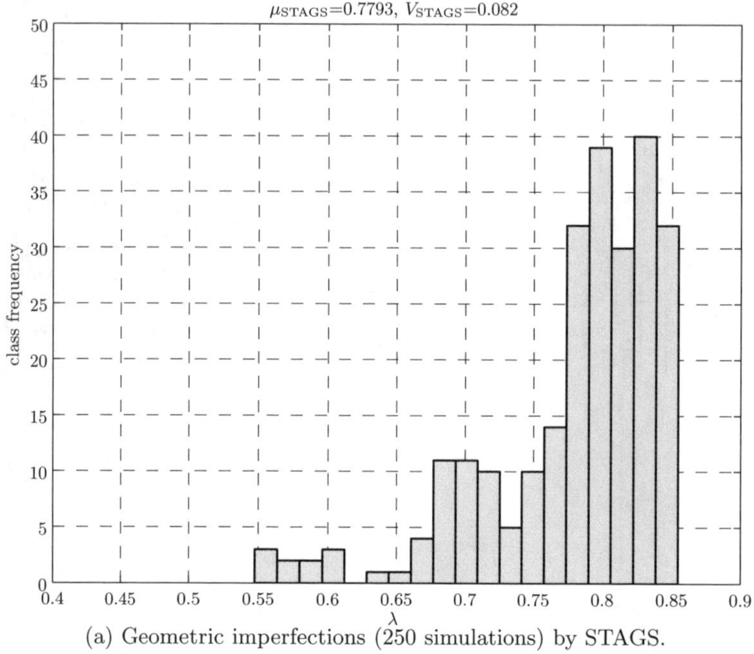

(a) Geometric imperfections (250 simulations) by STAGS.

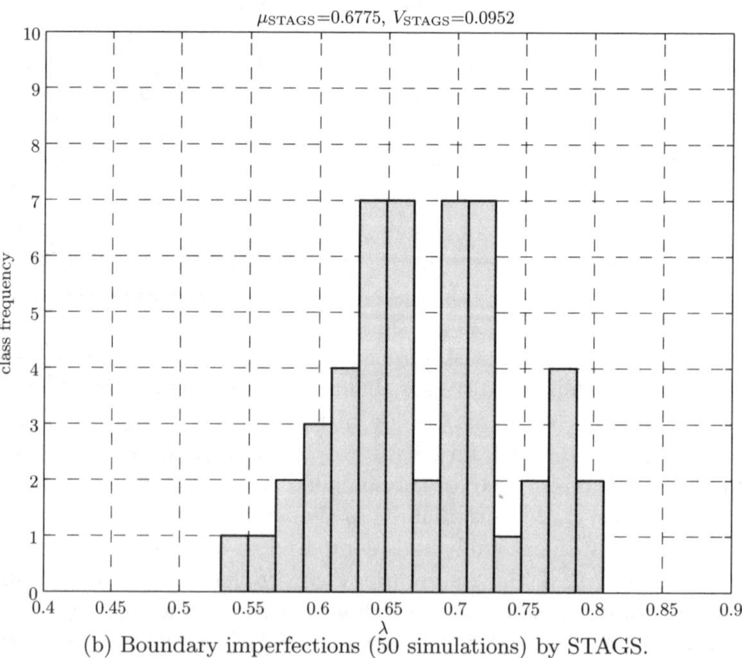

(b) Boundary imperfections (50 simulations) by STAGS.

Fig. 12.18. Histograms of the limit load for shell #1 (geometric or boundary imperfections).

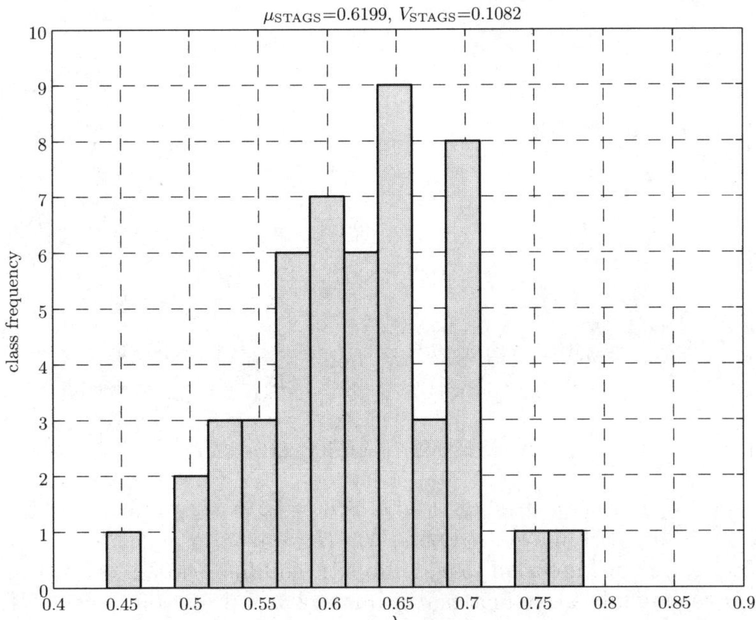

(a) Geometric and boundary imperfections (50 simulations) by STAGS.

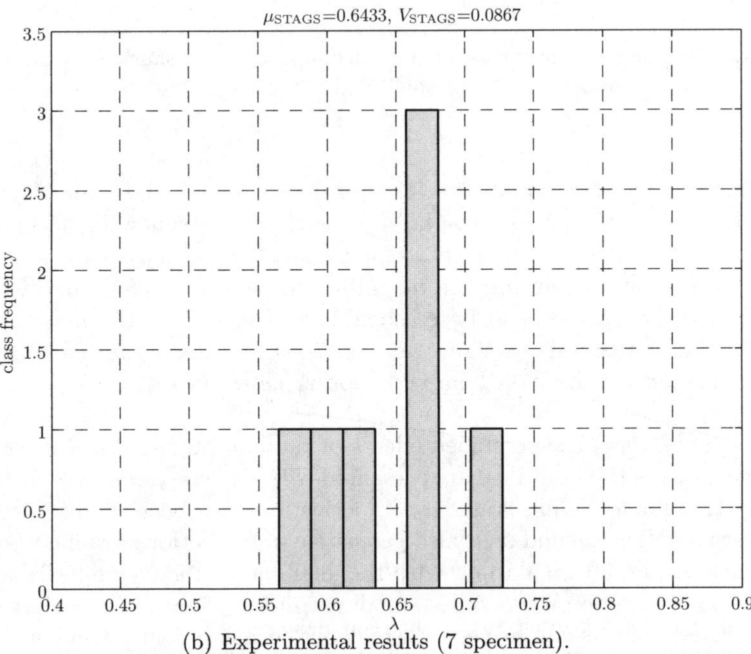

(b) Experimental results (7 specimen).

Fig. 12.19. Histograms of the limit load for shell #1 (geometric and boundary imperfections and experimental results).

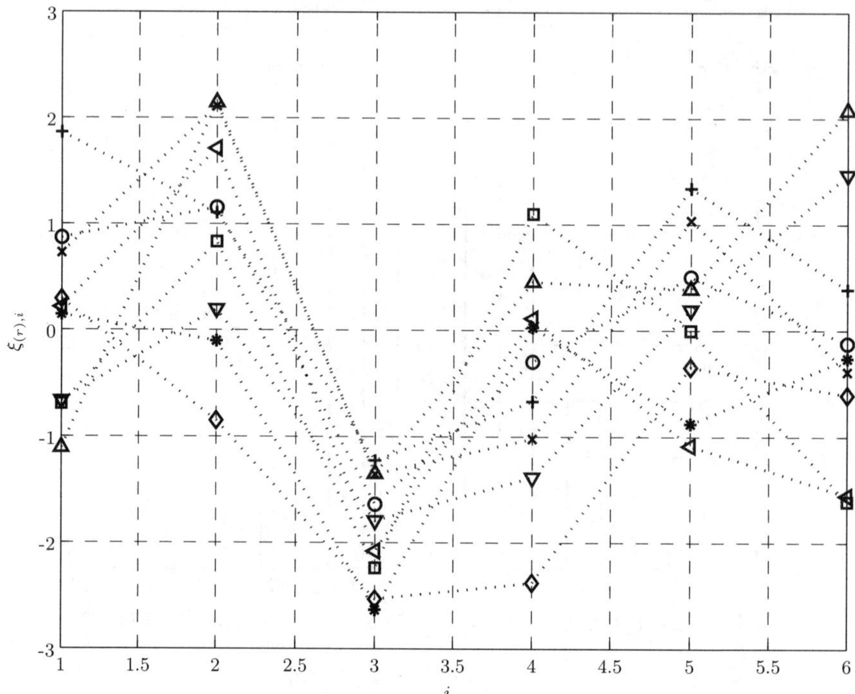

Fig. 12.20. Random variables ξ of the Karhunen-Loève expansion for most detrimental geometric imperfections for shell #1 ($\lambda_{\text{nl}}^{\text{lim}} < 0.6$).

this filling material was lower than that of the shell, with the result that there are still boundary imperfections, however with less detrimental effects on the limit load. In order to take into account this effect, a parameter study has been carried out by varying the magnitude of the boundary imperfections, revealing that it appears to be reasonable to use 15% of the magnitude of boundary imperfections as shown in Fig. A.5. The histogram of 50 simulations in presence of boundary imperfection is shown in Fig. 12.18(b).

In Fig. 12.19(a) the combined effects of random boundary and geometric imperfections of the limit load are presented. The two step approach as defined in Sec. 12.2 for modeling boundary imperfections has been modified in such a way, that in the second step also geometric imperfections are incorporated into the analysis. 50 simulations with the identical boundary imperfections as used in the simulation where only boundary imperfections are considered have been carried out. As can be seen, the calculated mean values and coefficients of variation do agree quite well with their experimentally determined counterparts. Considering both sources of imperfections – boundary and geometric ones – leads to a lower mean value and a larger coefficient of variation when compared with the results obtained for boundary imperfections only. This, of course, was to be expected.

12.4.3 Cylindrical Shell with Cutout and Random Geometric Imperfections

Geometric imperfections have been measured for very thin cylindrical shells ($R/t \approx 900$, see [5]), when compared with the shell used in this example ($R/t \approx 90$). In order to take this into account, the imperfection magnitude has been been normalized with respect to the shell thickness t. The maximum imperfection magnitude is about three times the shell thickness. The effect of three different magnitudes of geometric imperfections on the critical load have been investigated, i.e. $3.0\,t$, $1.5\,t$ and $0.3\,t$. According to the effect of the shape and magnitude of the generated imperfections on the stability behavior of the shell, different criteria of failure have been defined:

(i) If collapse occurs, e.g. no bifurcation point has been passed along the equilibrium path, then the critical load is defined to be the limit load.

(ii) If bifurcation occurs at a load level higher than the bifurcation load of the perfect shell $\lambda^{\mathrm{bif}}_{\mathrm{perf}}$ (corresponding to $\lambda^{\mathrm{bif}}_{\mathrm{nl}}$ in Table 12.2) and the load step before is lower than $\lambda^{\mathrm{bif}}_{\mathrm{perf}}$, then $\lambda^{\mathrm{bif}}_{\mathrm{perf}}$ is assumed to be the critical load. In this case the imperfections do have no significant influence on the buckling behavior of the shell, implying that the imperfect shell behaves similar to the perfect shell. In general, this would require an investigation of the secondary solution path, but this is for automated Monte Carlo simulation rather difficult to achieve. In this case however, the definition of the bifurcation load as the critical load yields a conservative approximation of the critical load.

(iii) If bifurcation occurs at a load level below or above the bifurcation load of the perfect shell, then the mean value of that load step and the load step before is assumed to be the critical load. It has to be noted, that the calculation of a bifurcation point is not as straight forward as the calculation of the limit load and requires in general a restart of the analysis. Again, for Monte Carlo simulation this is rather difficult to achieve.

Direct Monte Carlo simulation for two different cutout sizes and three different magnitudes of imperfections has been carried out. In particular, 50 simulations for each case, see Fig. 12.21(a) - 12.23(b), have been performed. The CPU time required for a non-linear analysis was in the range of 2,500 and 4,000 seconds on a SUN Solaris ULTRA 2/400, depending on the critical load level and the necessary number of iterations, governed by the shape and magnitude of the geometric imperfections. As can be seen from Fig. 12.21(a) - 12.23(b), the mean value is monotonically increasing with a decreasing imperfection magnitude for both sizes of cutouts. It has to be noted that for shell #2 and imperfection magnitudes $1.5\,t$ and $0.3\,t$ (see Fig. 12.22(a) and

Fig. 12.23(a), respectively), the mean value μ_{STAGS} of the critical load is higher than the bifurcation load of the perfect shell. This is not the case for shell #3, where for all cases μ_{STAGS} is lower than the bifurcation load of the perfect shell. The coefficient of variation V_{STAGS} of the critical load for shell #3 clearly decreases with decreasing imperfection magnitude. This tendency can not be observed for shell #2, where V_{STAGS} for imperfection magnitude $0.3\,t$ is larger than for $1.5\,t$.

12.5 Summarizing Remarks

At the end of this chapter, some important features of the described procedure for the stability analysis of large finite element systems with random imperfection are summarized.
Random geometric imperfections have been modeled according to the theory of random fields, therefore no idealization in terms of modal geometric imperfections has to be made, i.e. imperfections are modeled in a more realistic way. In this context, the Karhunen-Loève expansion, being capable to generate non-homogeneous geometric imperfections, turns out to be an appropriate tool in the context with Monte Carlo simulation.
Random boundary imperfections have been modeled as a stationary process. In this case, too, no idealization in terms of modal boundary imperfections is therefore necessary.
The Monte Carlo simulation technique has been successfully applied in predicting the scatter in the limit load of axially compressed cylindrical shells, as it is observed in experiments. Due to the extensive modeling capabilities of the finite element method, this methodology can be extended to a more complex stability analysis, e.g. with additional sources of non-linearities as well as complex shell structures. This attractive feature has been demonstrated by applying the method to cylindrical shells with a cutout and random geometric imperfections.
A minor drawback of this approach is that the finite element mesh for thin shell structures has to be rather fine in order to reproduce the shell response sufficiently well. Hence this approach is computationally expensive. However, it is the authors' belief that the fast and continuing evolution of digital computers will compensate for this drawback already in the immediate future.
A database of measured imperfections, for instance for shells structures manufactured through a certain production process, should be available in order to fully exploit the capabilities of this approach. Using the method as presented in this monograph enables one to assess the critical load in a more realistic way, hence providing the analyst with a tool for predicting the critical load in terms of its second moment characteristics. This most valuable information can be used for the theoretical verification of design recommendations for imperfection sensitive shell structures, as shown for cylindrical shells with and without cutouts under axial compression.

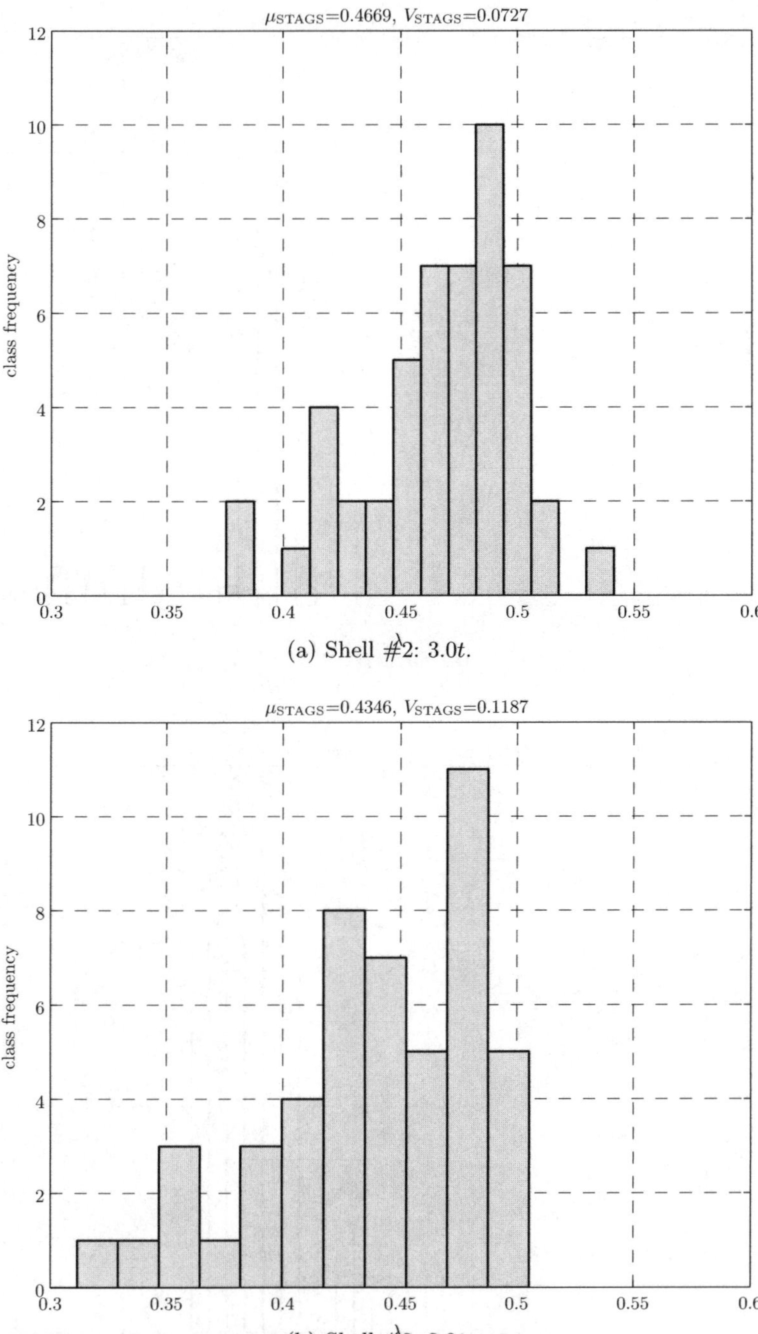

(a) Shell #2: 3.0t.

(b) Shell #3: 3.0t.

Fig. 12.21. Histograms of the limit load for shell #2 and shell #3 (3.0t).

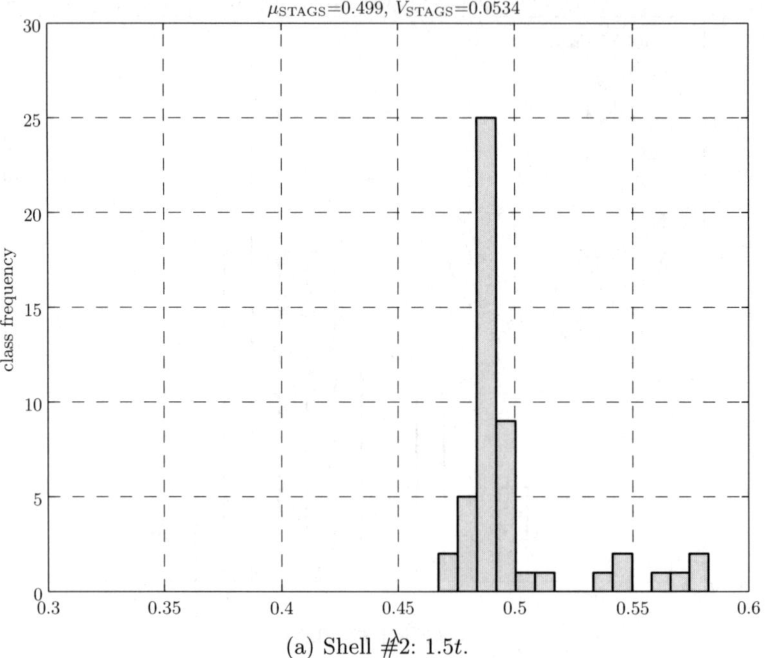

(a) Shell #2: 1.5t.

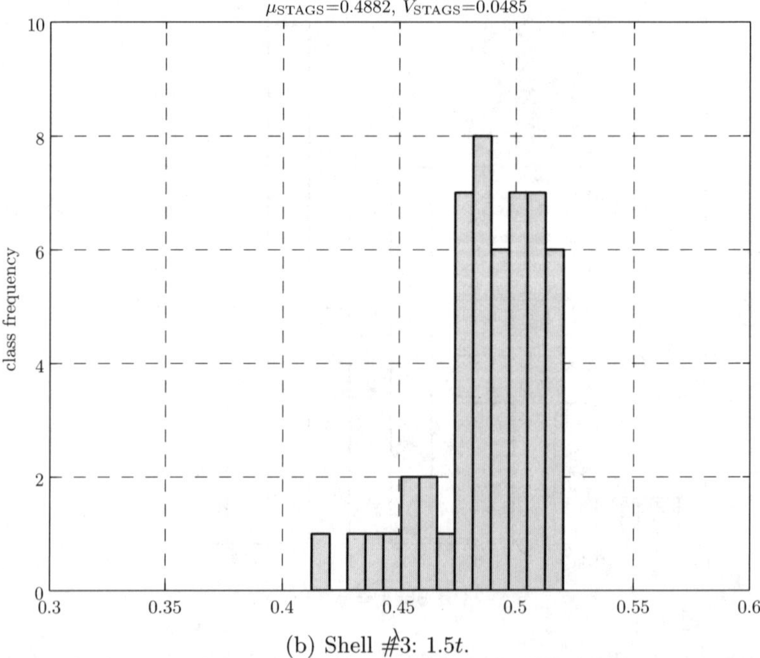

(b) Shell #3: 1.5t.

Fig. 12.22. Histograms of the limit load for shell #2 and shell #3 (1.5t).

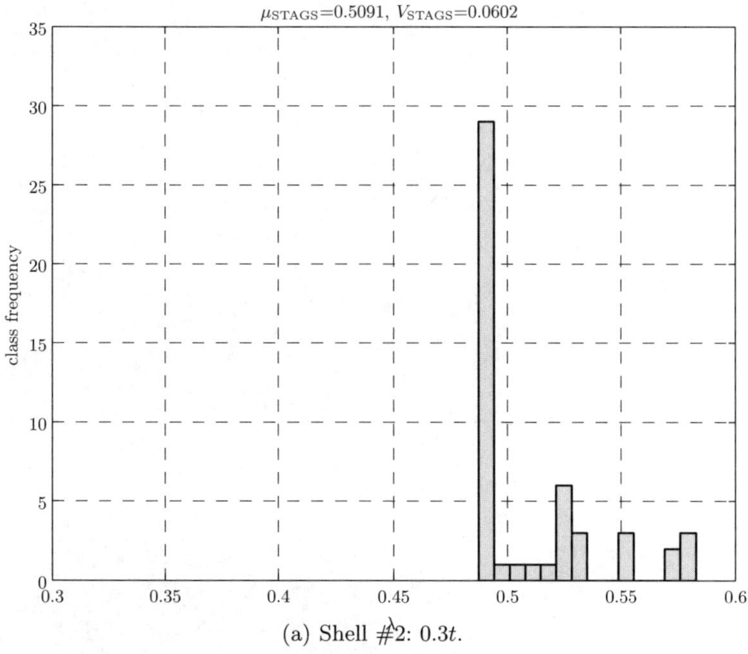

(a) Shell #2: 0.3t.

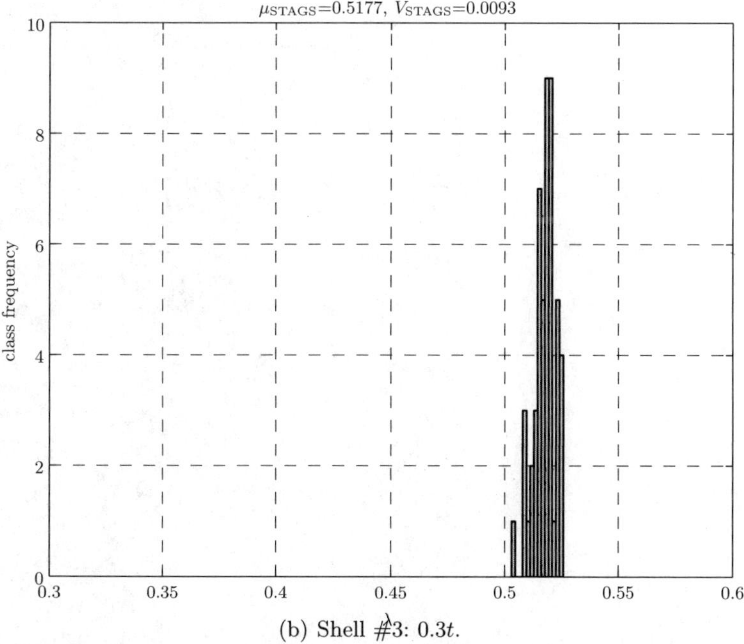

(b) Shell #3: 0.3t.

Fig. 12.23. Histograms of the limit load for shell #2 and shell #3 (0.3t).

13

Random Vibrations of Multi-Story Structures

The procedure described in Sec.11 is now applied to both a multi degree of freedom system (12-story building, 36 degrees of freedom) and a large finite element system (6-story office building with 24,984 degrees of freedom). In all these numerical examples, non-linearity is modeled by hysteretic elements as described in Sec. 13.2. Because of the need to represent filtered white noise by the Karhunen-Loève expansion in these numerical examples, a rather large number of Karhunen-Loève vectors is required in order to keep the truncation error acceptably small, which, in fact, may not be true for modeling measured excitation processes. Finally, it is quite obvious, that the computational effort of the proposed approach correlates strongly with the number of Karhunen-Loève vectors used.

13.1 Stochastic Modeling of Earthquake Excitation

As in many cases, due to the lack of experimental data and other pertinent geotectonic information, the excitation for both systems will be modeled as non-stationary, Gaussian, filtered white noise. The Karhunen-Loève representation is valid for any type of non-white excitation and can be easily determined if second moment properties of the underlying process are known.

13.1.1 Karhunen-Loève Representation of Filtered White Noise

The components $a(t)$ of the vector of accelerations $\boldsymbol{a}(t)$ are modeled as statistically independent stochastic processes defined as filtered white noise, i.e.

$$a(t) = \boldsymbol{Q}_f \boldsymbol{v}(t) , \tag{13.1}$$

where $\boldsymbol{Q}_f$ denotes a constant matrix and the filter follows the first order differential equation

$$\dot{\boldsymbol{v}}(t) = \boldsymbol{A}_f \boldsymbol{v}(t) + \boldsymbol{b}_f(t)\boldsymbol{w}(t) , \tag{13.2}$$

see also (B.11). In this context, $\boldsymbol{v}$, $\boldsymbol{A}_f$ and $\boldsymbol{b}_f$ denote the state vector, the system matrix and the so-called distribution matrix of the filter, respectively, and $\boldsymbol{w}(t)$ denotes the white noise vector as defined in (B.18) and (B.20). The covariance matrix $\boldsymbol{\Gamma}_{\boldsymbol{vv}}(t)$, describing the correlation of the filter parameters for a specific time t, can be calculated using the Lyapunov matrix differential equation, see also (B.22),

$$\dot{\boldsymbol{\Gamma}}_{\boldsymbol{vv}}(t) = \boldsymbol{A}_f \boldsymbol{\Gamma}_{\boldsymbol{vv}}(t) + \boldsymbol{\Gamma}_{\boldsymbol{vv}}(t)\boldsymbol{A}_f^T + \boldsymbol{b}_f(t)\boldsymbol{I}_w(t)\boldsymbol{b}_f^T(t) . \tag{13.3}$$

In addition, the covariance matrix $\boldsymbol{\Gamma}_{\boldsymbol{vv}}(t,s)$, describing the correlation of the filter parameters between the two time instances t and s, can be calculated according to the matrix differential equation, see e.g. [141],

$$\frac{\partial}{\partial t}\boldsymbol{\Gamma}_{\boldsymbol{vv}}(t,s) = \boldsymbol{A}_f \boldsymbol{\Gamma}_{\boldsymbol{vv}}(t,s) , \tag{13.4}$$

where

$$\boldsymbol{\Gamma}_{\boldsymbol{vv}}(s,s) = \boldsymbol{\Gamma}_{\boldsymbol{vv}}(s) . \tag{13.5}$$

The covariance function $\Gamma_{aa}(t,s)$ can then, using (13.1), be calculated by

$$\Gamma_{aa}(t,s) = E\{\boldsymbol{Q}_f \boldsymbol{v}(t)\boldsymbol{v}^T(s)\boldsymbol{Q}_f^T\} . \tag{13.6}$$

By solving the algebraic eigenvalue problem of the matrix $\boldsymbol{\Gamma}_{aa}(t,s)$ – obtained by discretizing $\Gamma_{aa}(t,s)$ – as stated in (8.14), the Karhunen-Loève representation of the filtered white noise process $a(t)$ can then be determined.
The power spectral density $S_{aa}(\omega)$ of the stationary ground acceleration captures some of the excitation characteristics of the non-stationary ground acceleration and provides information on the frequency content of the excitation process. It can be obtained by the Fourier transformation of the stationary form of (13.6), yielding

$$S_{aa}(\omega) = \boldsymbol{Q}_f \boldsymbol{S}_{\boldsymbol{vv}}(\omega)\boldsymbol{Q}_f^T . \tag{13.7}$$

For this purpose, the power spectral density matrix $\boldsymbol{S}_{\boldsymbol{vv}}(\omega)$ of the stationary filter outputs defined by, see (13.14),

$$\boldsymbol{b}_f = \begin{Bmatrix} 0 \\ 1 \\ 0 \\ 0 \end{Bmatrix} , \tag{13.8}$$

has to be determined. This can be done by using the relation

$$\boldsymbol{S}_{\boldsymbol{vv}}(\omega) = \boldsymbol{H}^*(\omega) S_{\boldsymbol{ww}}(\omega)\boldsymbol{H}^T(\omega) = \boldsymbol{H}^*(\omega)\frac{1}{2\pi}\boldsymbol{b}_f \boldsymbol{I}_{\boldsymbol{w}} \boldsymbol{b}_f^T \boldsymbol{H}^T(\omega) , \tag{13.9}$$

where the stationary frequency response function $\boldsymbol{H}(\omega)$ is given by

$$\boldsymbol{H}(\omega) = (-\boldsymbol{A}_f + j\omega \boldsymbol{I})^{-1} , \tag{13.10}$$

$j = \sqrt{-1}$ and the superscript '*' indicates complex conjugate. The frequently used one-sided power spectral density $G_{aa}(\omega)$ can be determined according to

$$G_{aa}(\omega) = 2S_{aa}(\omega) , \qquad 0 \leq \omega \leq \infty . \tag{13.11}$$

Using this methodology, the Karhunen-Loève representation of filtered white noise can calculated efficiently. It should be stressed, however, that the Karhunen-Loève expansion is a much more powerful tool in representing non-stationary second moment properties of stochastic processes when compared with linearly filtered white noise. The covariance matrix of a stochastic process can capture by definition non-stationary processes, while there is no unique definition of the associated time-varying power spectrum [163]. The theory of evolutionary spectra [101] is capable describing the second moment characteristics of a non-stationary stochastic process, its implementation in random vibrations is, however, rather involved.

13.1.2 Filtered White Noise Model

With respect to (13.1), it is assumed that

$$\boldsymbol{Q}_f = \left\{ \Omega_{1g}^2 \; 0 \; -\Omega_{2g}^2 \; -2\zeta_{2g}\Omega_{2g} \right\} , \tag{13.12}$$

$$\boldsymbol{A}_f = \begin{bmatrix} 0 & 1 & 0 & 0 \\ -\Omega_{1g}^2 & -2\zeta_{1g}\Omega_{1g} & 0 & 0 \\ 0 & 0 & 0 & 1 \\ \Omega_{1g}^2 & 2\zeta_{1g}\Omega_{1g} & -\Omega_{2g}^2 & -2\zeta_{2g}\Omega_{2g} \end{bmatrix} , \tag{13.13}$$

and

$$\boldsymbol{b}_f = \begin{Bmatrix} 0 \\ e(t) \\ 0 \\ 0 \end{Bmatrix} , \tag{13.14}$$

where the time modulation function $e(t)$ is assumed to

$$e(t) = \frac{\mathrm{e}^{-at} - \mathrm{e}^{-bt}}{\max(\mathrm{e}^{-at} - \mathrm{e}^{-bt})} . \tag{13.15}$$

13.1.3 Comparison Between Karhunen-Loève Representation and Filtered White Noise

In this section, the equivalence of modeling a second order process by either its Karhunen-Loève expansion or by filtered white noise is discussed. In particular, the accuracy of the Karhunen-Loève representation with regard to the order of truncation of the Karhunen-Loève expansion is addressed. While used here in a different context, it is also possible to apply the Karhunen-Loève representation of filtered white noise for an effective reduction of dimensionality in reliability analysis of dynamical systems, where quite large number of approaches rely on white noise input.

The covariance matrix of a component of the ground acceleration $\boldsymbol{\Gamma}_{aa}(t,s)$ has been obtained by (13.6), where the covariance matrix $\boldsymbol{\Gamma}_{\boldsymbol{vv}}(t,s)$ of the filter output between two time instances has been determined according to (13.3) and (13.4) by a Runge-Kutta algorithm.

A time span of [0,20] seconds has been integrated with a time step $\Delta t = 0.01$ seconds, yielding a size 2001×2001 for $\boldsymbol{\Gamma}_{aa}(t,s)$. The values $\Omega_{1g} = 15.0$ rad/s, $\zeta_{1g} = 0.8$, $\Omega_{2g} = 0.3$ rad/s, $\zeta_{2g} = 0.995$, and the white noise intensity $I = 0.18$ m^2/s^3 have been used for the filter system matrix (13.13) and the white noise loading process.

The eigenvalues $\{\lambda_i\}_{i=1}^{1000}$ of $\boldsymbol{\Gamma}_{aa}(t,s)$ are shown in Fig. 13.1 in a semi-logarithmic (base 10) scale. By definition, $\boldsymbol{\Gamma}_{aa}(t,s)$ is a real self adjoint matrix

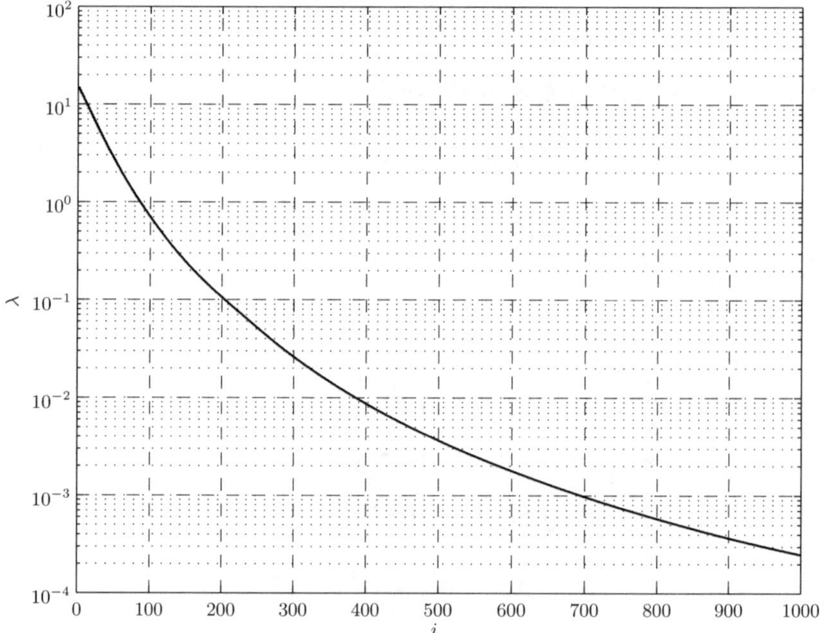

Fig. 13.1. Eigenvalues $\{\lambda_i\}_{i=1}^{1000}$ of the covariance matrix $\boldsymbol{\Gamma}_{aa}(t,s)$.

operator, thus all eigenvalues and eigenvectors are real. As can be seen from Fig. 13.1, the eigenvalues λ_i decrease rather quickly for increasing i. The eigenvectors $\{\boldsymbol{\psi}_i(t)\}_{i=1}^{4}$ are plotted in Fig. 13.2 for the time span of [0,10] seconds.

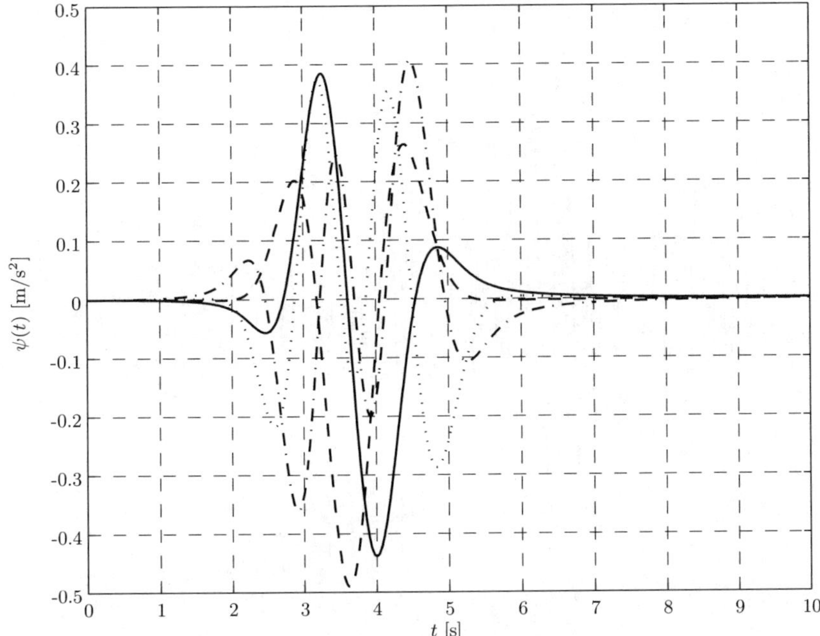

Fig. 13.2. Eigenvectors $\psi(t)$ of the covariance matrix $\boldsymbol{\Gamma}_{aa}(t,s)$; $i = 1$ (solid), $i = 2$ (dashed), $i = 3$ (dashdot) and $i = 4$ (dotted).

For this particular kind of excitation, the lower order eigenvectors are acting localized with regard to the integration time span of 20 seconds. A realization of the stochastic ground acceleration $a(t)$ using 300 Karhunen-Loève vectors is shown in Fig. 13.3.
In a next step, the variances of the response of a simple single degree of freedom system are compared – obtained by either the Karhunen-Loève representation or the filtered white noise model for the ground acceleration process, respectively. In case of the Karhunen-Loève representation of the ground accelerations, the approach as described Chap. 11 is applied. The equation of motion of a linear single degree of freedom system is given by

$$M\ddot{u} + C\dot{u} + Ku = f(t) , \tag{13.16}$$

where the parameters follow the nomenclature of (6.18) and are assumed to $M = 1$ kg, $C = 0.6$ kgs^{-1}, $K = 100$ kgs^{-2}. In accordance to (11.1), the external force is taken to $f(t) = -Ma(t)$. The state space equation is given by

$$\dot{\boldsymbol{x}} = \boldsymbol{A}_s\boldsymbol{x} + \boldsymbol{b}_s f(t) , \tag{13.17}$$

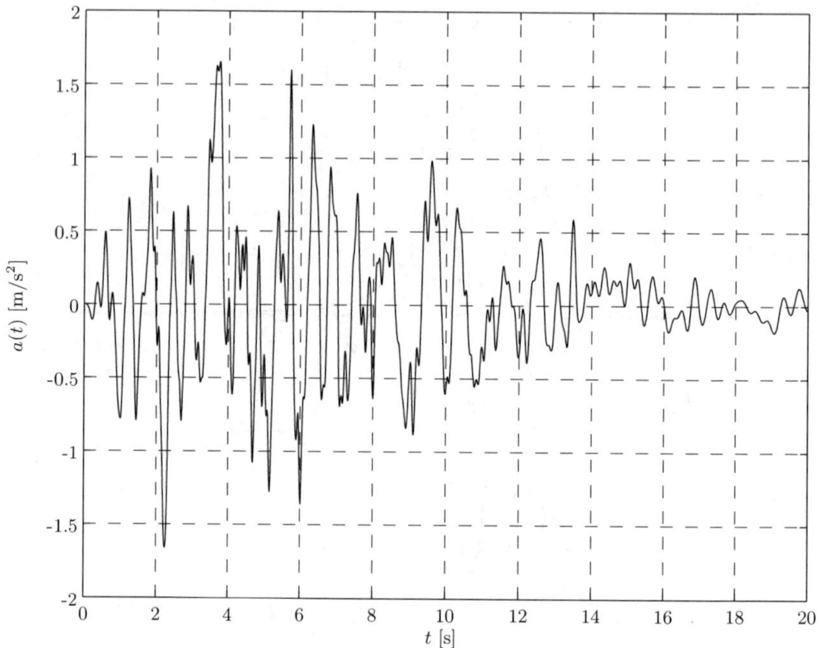

Fig. 13.3. Realization of the stochastic ground acceleration $a(t)$ using $m = 300$ Karhunen-Loève vectors.

where

$$\boldsymbol{x} = \begin{Bmatrix} u \\ \dot{u} \end{Bmatrix}, \qquad \boldsymbol{b}_s(t) = \begin{Bmatrix} 0 \\ M^{-1} \end{Bmatrix}, \tag{13.18}$$

and

$$\boldsymbol{A}_s = \begin{bmatrix} 0 & 1 \\ -M^{-1}K & -M^{-1}C \end{bmatrix}. \tag{13.19}$$

The state space equation of the combined system (single degree of freedom system system and filter), i.e. (13.16), (13.1) and (13.2) is given by

$$\dot{\boldsymbol{y}} = \boldsymbol{A}\boldsymbol{y} + \boldsymbol{b}(t)\boldsymbol{w}(t), \tag{13.20}$$

where

$$\boldsymbol{y} = \begin{Bmatrix} \boldsymbol{x} \\ \boldsymbol{v} \end{Bmatrix}, \qquad \boldsymbol{b}(t) = \begin{Bmatrix} \boldsymbol{0} \\ \boldsymbol{b}_f \end{Bmatrix}, \tag{13.21}$$

and

$$\boldsymbol{A} = \begin{bmatrix} \boldsymbol{A}_s & -\boldsymbol{b}_s\boldsymbol{Q} \\ \boldsymbol{0} & \boldsymbol{A}_f \end{bmatrix}. \tag{13.22}$$

By solving the Lyapunov matrix differential equation of the combined system, see also (B.22),

$$\dot{\boldsymbol{\Gamma}}_{\boldsymbol{yy}}(t) = \boldsymbol{A}\boldsymbol{\Gamma}_{\boldsymbol{yy}}(t) + \boldsymbol{\Gamma}_{\boldsymbol{yy}}(t)\boldsymbol{A}^T + \boldsymbol{b}(t)\boldsymbol{I}_w\boldsymbol{b}^T(t) . \tag{13.23}$$

and by solving (13.16) for every Karhunen-Loève vector $a^{(j)}(t)$, i.e.

$$\dot{\boldsymbol{x}}^{(j)} = \boldsymbol{A}_s\boldsymbol{x}^{(j)} - \boldsymbol{b}_s M a^{(j)}(t) , \tag{13.24}$$

it is possible verify the error in the variance function of the response due to truncation of the Karhunen-Loève expansion.

When integrating the Karhunen-Loève vectors according to (13.24), their contribution after a certain time to the overall response is negligible and could be omitted. The support of the higher order eigenvectors is less localized, see Fig. 13.4, but due to the weighting of each eigenvector with the corresponding eigenvalue, $\boldsymbol{x}^{(j)}(t) = \sqrt{\lambda_j}\boldsymbol{\psi}_j(t)$, their contribution to the overall response decreases for increasing j. This allows to truncate the series expansion at order m.

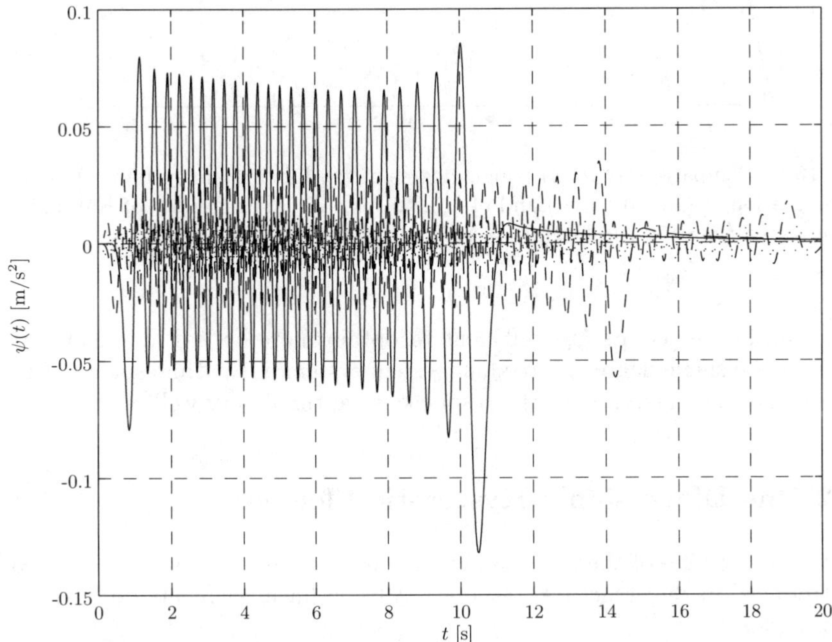

Fig. 13.4. Eigenvectors $\psi(t)$ of the covariance matrix $\boldsymbol{\Gamma}_{aa}(t,s)$; $i = 50$ (solid), $i = 100$ (dashed), $i = 200$ (dashdot) and $i = 300$ (dotted).

In Fig. 13.5 the variance function of the single degree of freedom system system obtained for different values of m of the Karhunen-Loève expansion compared to the filtered white noise loading is shown. According to these results, 100

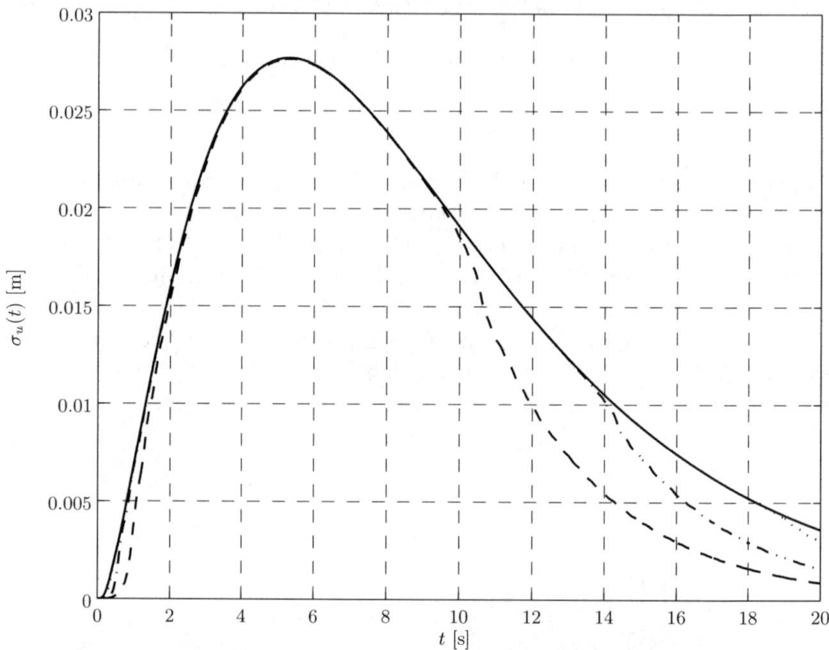

Fig. 13.5. Comparison of $\sigma_u(t)$ (single degree of freedom system system) obtained by integration of (13.24) (solid) and of (13.23) for a different number of Karhunen-Loève vectors: $m = 50$ (dashed), $m = 100$ (dashdot), $m = 200$ (dotted).

Karhunen-Loève vectors are sufficient to obtain more or less identical variances for the single degree of freedom system system in the time span [0.5,14] seconds, i.e. the maximum of the response is captured very well.

13.2 One Dimensional Hysteretic Elements

Due to the yielding of the material, energy dissipation, as shown in Fig. 13.6, is introduced in the structural response. All non-linear elements do have the restoring force

$$r_L^{(e)} = k^{(e)}(u_L^{(e)} - (q_1^{(e)} + q_2^{(e)})) = k^{(e)}u_L^{(e)} + \bar{r}_L^{(e)} , \tag{13.25}$$

where $k^{(e)}$ denotes the initial stiffness of the non-linear element, $u_L^{(e)}$ the relative displacement between the two degrees of freedom to which the element is connected. By building the structural matrices, the initial stiffness of the non-linear elements has been already considered in order to keep the right hand side of the non-linear system equation (6.4) small. Hence the force arising from a non-linear element reads, see also (11.20),

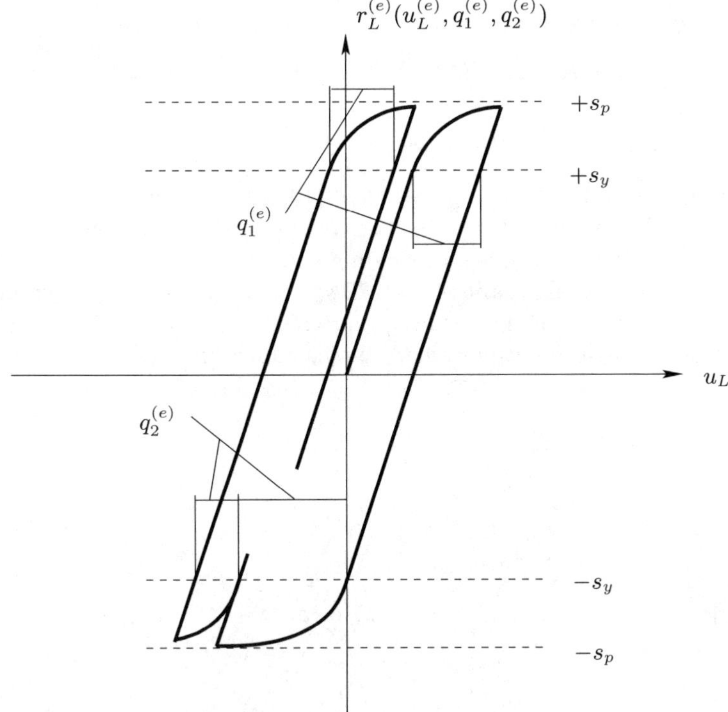

Fig. 13.6. Hysteretic material behavior.

$$\bar{r}_L^{(e)} = -\boldsymbol{R}_L^{(e)} \boldsymbol{q}^{(e)} = -k^{(e)}(q_1^{(e)} + q_2^{(e)}) \,, \tag{13.26}$$

where

$$\boldsymbol{R}_L^{(e)} = \left\{ k^{(e)}, k^{(e)} \right\} \,, \tag{13.27}$$

and

$$\boldsymbol{q}^{(e)} = \begin{Bmatrix} q_1^{(e)} \\ q_2^{(e)} \end{Bmatrix} \tag{13.28}$$

is a vector containing the auxiliary variables, which in turn represent plastic elongations. Introducing the additional variables

$$x_1^{(e)} = u_L^{(e)} - (q_1^{(e)} + q_2^{(e)}) \,, \qquad x_2^{(e)} = \dot{u}_L^{(e)} \,, \tag{13.29}$$

the plastic elongations are specified by the non-linear differential equations

$$\begin{aligned} \dot{q}_1 &= g_1(x_1, x_2) \\ &= x_2 H(x_2) \cdot [\; H(x_1 - s_y) \frac{x_1 - s_y}{s_p - s_y} H(s_p - x_1) + H(x_1 - s_p) \;] \,, \end{aligned} \tag{13.30}$$

and

$$\dot{q}_2 = g_2(x_1, x_2)$$
$$= -x_2 H(-x_2) \cdot \left[\, H(-x_1 - s_y) \frac{-x_1 - s_y}{s_p - s_y} H(s_p + x_1) + H(-x_1 - s_p) \,\right], \tag{13.31}$$

where the superscript "$^{(e)}$" in (13.30) and (13.31) has been dropped for better readability. The symbol H denotes the Heaviside step function, s_y is a given parameter specifying the onset of yielding and $k \cdot s_p$ is the maximum restoring force of the element. The value $s_y = 0.70\ s_p$ has been assuem for all non-linear elements. It should be mentioned that variables q_1 and q_2 provide information on the total plastic deformation. A typical function $g_1(x_1, x_2)$ is shown in Fig. 13.7.

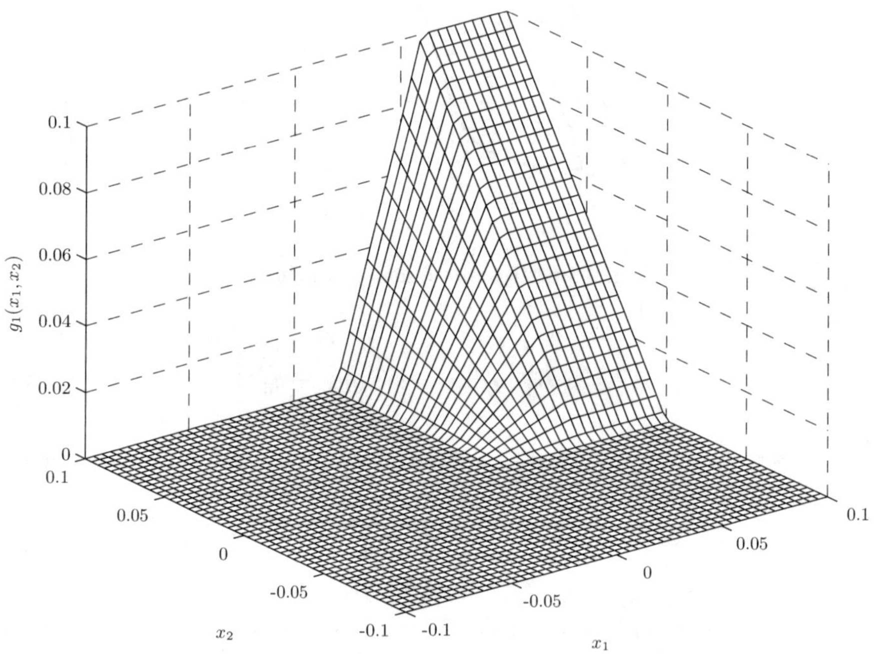

Fig. 13.7. Typical non-linear function $g_1(x_1, x_2)$; ($s_p = 0.06$, $s_y = 0.02$).

The two dimensional Gaussian probability density function for the variables introduced in (13.29) reads

$$p(x_1, x_2) = \frac{1}{2\pi\sigma_1\sigma_2\sqrt{1-\rho^2}} \exp\Big\{ -\frac{1}{2(1-\rho^2)}\Big[\Big(\frac{x_1-\mu_1}{\sigma_1}\Big)^2 - 2\rho\Big(\frac{x_1-\mu_1}{\sigma_1}\Big)\Big(\frac{x_2-\mu_2}{\sigma_2}\Big) + \Big(\frac{x_2-\mu_2}{\sigma_2}\Big)^2\Big]\Big\}, \tag{13.32}$$

where μ_1 and μ_2 denotes the mean, σ_1 and σ_2 the standard deviation of the random variables x_1 and x_2, respectively, and ρ the correlation coefficient, see Fig. 13.8. The non-linear function (13.30) is replaced by the linear relation

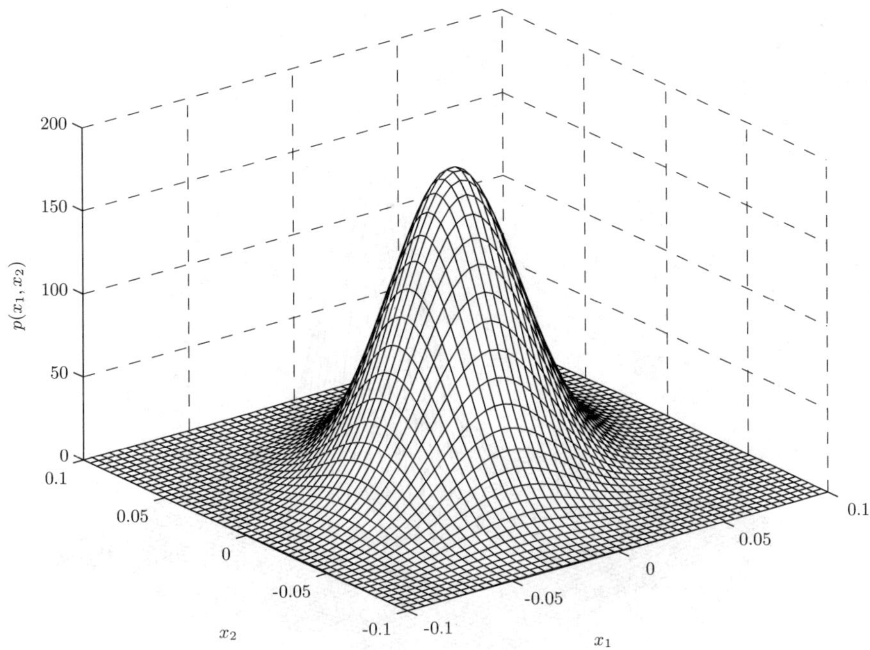

Fig. 13.8. Typical 2-dimensional Gaussian density $p(x_1, x_2)$; ($\mu_1 = \mu_2 = 0$, $\sigma_1 = \sigma_2 = 0.03$, $\rho = 0.3$).

$$\dot{q}_1 = b_{q_1} + c_{q_1,1}(x_1 - E\{x_1\}) + c_{q_1,2}(x_2 - E\{x_2\}) , \tag{13.33}$$

where, in accordance to [9], the linearization coefficients can be determined by the integrals

$$b_{q_1} = E\{g_1(x_1, x_2)\} = \int_{x_1=s_y}^{\infty} \int_{x_2=0}^{\infty} g_1(x_1, x_2) p(x_1, x_2)\, \mathrm{d}x_1 \mathrm{d}x_2 \tag{13.34}$$

and

$$c_{q_1,1} = E\Big\{\frac{\partial g_1(x_1,x_2)}{\partial x_1}\Big\} = \int_{x_1=s_y}^{s_p}\int_{x_2=0}^{\infty}\frac{x_2}{s_p - s_y}\,p(x_1,x_2)\,\mathrm{d}x_1\mathrm{d}x_2\;, \qquad (13.35)$$

$$c_{q_1,2} = E\Big\{\frac{\partial g_1(x_1,x_2)}{\partial x_2}\Big\} =$$
$$\int_{x_1=s_y}^{\infty}\int_{x_2=0}^{\infty}\Big[\frac{x_1 - s_y}{s_p - s_y}H(s_p - x_1) + H(x_1 - s_p)\Big]p(x_1,x_2)\,\mathrm{d}x_1\mathrm{d}x_2\;. \qquad (13.36)$$

The integrands of (13.34)-(13.36) are shown in Fig. 13.9 - 13.11. In the same

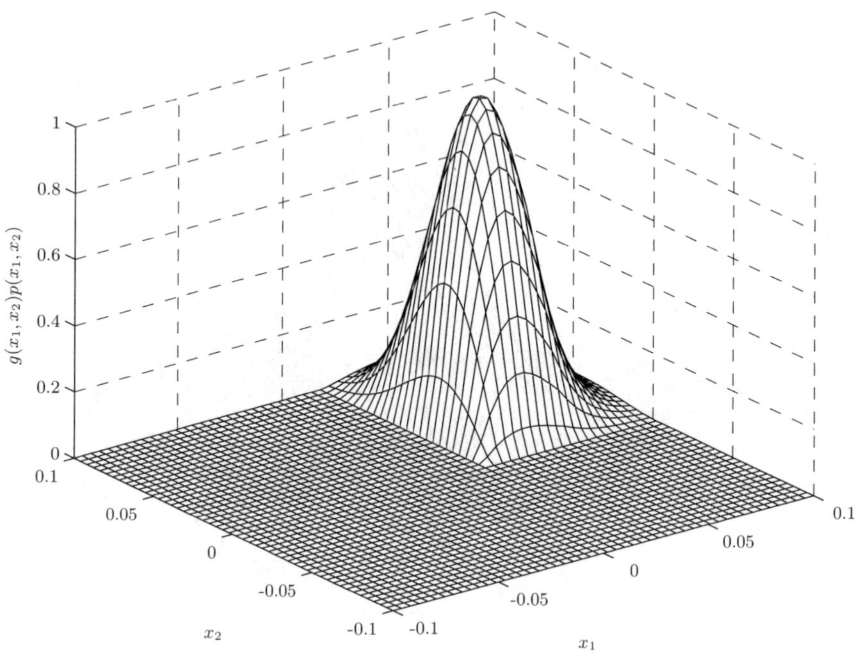

Fig. 13.9. Typical integrand of (13.34); ($s_p = 0.06$, $s_y = 0.02$, $\mu_1 = \mu_2 = 0$, $\sigma_1 = \sigma_2 = 0.03$, $\rho = 0.3$).

way, the non-linear function (13.31) is replaced by the linear relation

$$\dot{q}_2 = b_{q_2} + c_{q_2,1}(x_1 - E\{x_1\}) + c_{q_2,2}(x_2 - E\{x_2\})\;, \qquad (13.37)$$

and the linearization coefficients can be determined by the integrals

$$b_{q_2} = E\{g_2(x_1,x_2)\} = \int_{x_1=s_y}^{\infty}\int_{x_2=0}^{\infty} g_2(x_1,x_2)p(x_1,x_2)\,\mathrm{d}x_1\mathrm{d}x_2\;, \qquad (13.38)$$

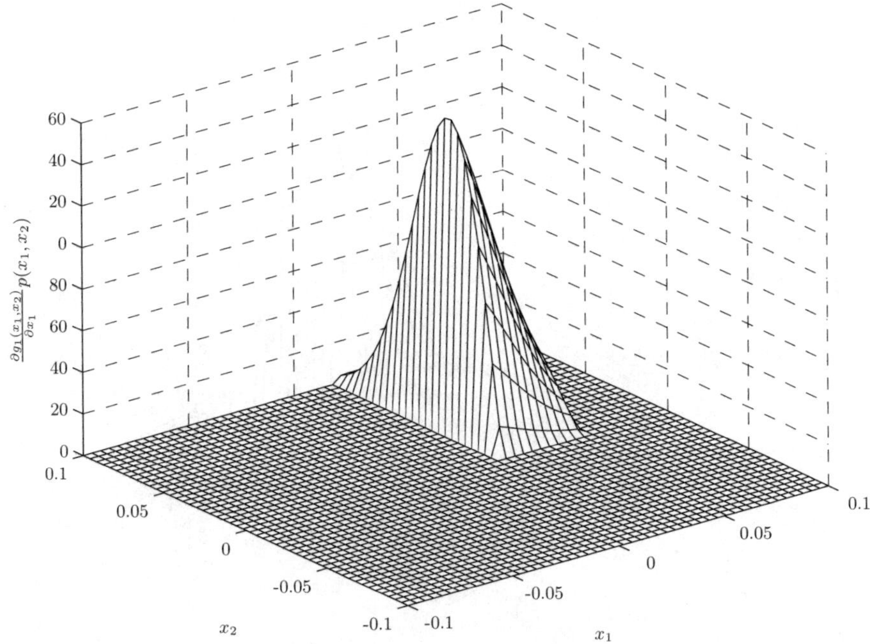

Fig. 13.10. Typical integrand of (13.35); ($s_p = 0.06$, $s_y = 0.02$, $\mu_1 = \mu_2 = 0$, $\sigma_1 = \sigma_2 = 0.03$, $\rho = 0.3$).

and

$$c_{q2,1} = E\Big\{\frac{\partial g_2(x_1,x_2)}{\partial x_1}\Big\} = \int_{x_1=s_y}^{s_p}\int_{x_2=0}^{\infty} \frac{x_2}{s_p - s_y}\, p(x_1,x_2)\,\mathrm{d}x_1\mathrm{d}x_2 \,, \quad (13.39)$$

$$c_{q2,2} = E\Big\{\frac{\partial g_2(x_1,x_2)}{\partial x_2}\Big\} =$$
$$\int_{x_1=s_y}^{\infty}\int_{x_2=0}^{\infty} \Big[\frac{x_1 + s_y}{s_p - s_y} H(s_p + x_1) - H(-x_1 - s_p)\Big] p(x_1,x_2)\,\mathrm{d}x_1\mathrm{d}x_2 \,. \quad (13.40)$$

Equations (13.32)-(13.40) refer to the element level, hence the superscript "(e)" has been dropped for better readability.

13.3 Modal Lyapunov Matrix Differential Equation

In the following, results obtained by the modal Lyapunov matrix differential equation, see also Appendix B, are used as a semi-analytical reference solution for linear systems.

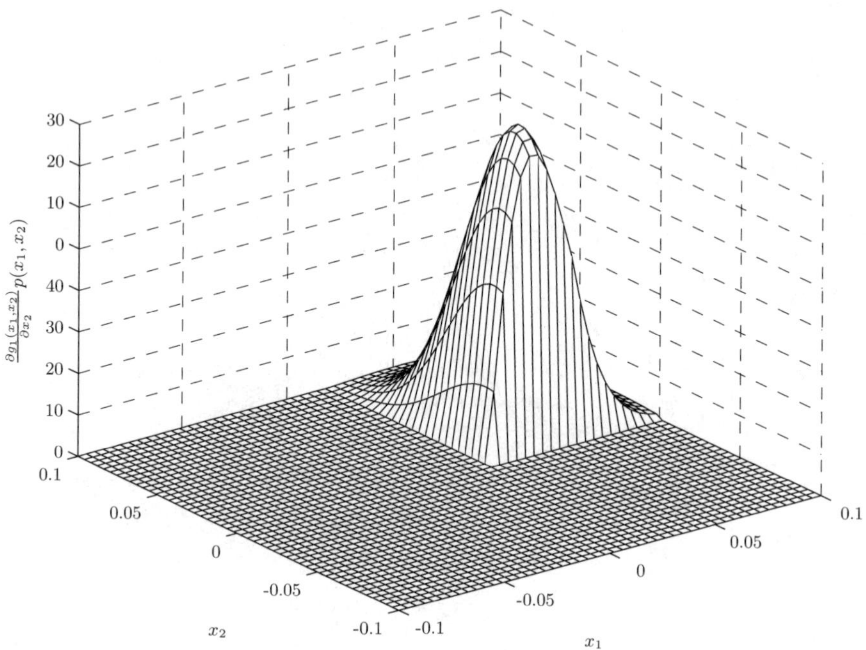

Fig. 13.11. Typical integrand of (13.36); ($s_p = 0.06$, $s_y = 0.02$, $\mu_1 = \mu_2 = 0$, $\sigma_1 = \sigma_2 = 0.03$, $\rho = 0.3$).

The mode displacement method as reviewed in Sec. 6.3 can also be applied to transform the Lyapunov matrix differential equation (B.22) into a reduced subspace, in which the temporal evolution of the covariance matrix $\boldsymbol{\Gamma}(t)$ can be solved very efficiently. The state vector and the loading vector, respectively, are in this case defined in terms of modal coordinates $\boldsymbol{u} = \boldsymbol{\Phi z}$ (compare with first part of (B.2)), i.e.

$$\boldsymbol{y} = \begin{Bmatrix} \boldsymbol{z} \\ \dot{\boldsymbol{z}} \end{Bmatrix}, \qquad \boldsymbol{b}(t) = \begin{Bmatrix} \boldsymbol{0} \\ \boldsymbol{\Phi}^T \boldsymbol{f}(t) \end{Bmatrix} \tag{13.41}$$

with

$$\boldsymbol{\Phi} = [\boldsymbol{\phi}_1, \boldsymbol{\phi}_2, \ldots, \boldsymbol{\phi}_N], \tag{13.42}$$

where N is the number of modes used. The system matrix for proportional damping is given by

$$\boldsymbol{A} = \begin{bmatrix} \boldsymbol{0} & \boldsymbol{I} \\ -\mathrm{diag}(\omega_i^2) & -\mathrm{diag}(2\zeta_i\omega_i) \end{bmatrix}, \tag{13.43}$$

where

$$\mathrm{diag}(\omega_i^2) := \begin{bmatrix} \omega_1^2 & 0 & \dots & 0 \\ 0 & \omega_2^2 & \ddots & \vdots \\ \vdots & \ddots & \ddots & 0 \\ 0 & \dots & 0 & \omega_N^2 \end{bmatrix} \tag{13.44}$$

and

$$\mathrm{diag}(2\zeta_i\omega_i) := \begin{bmatrix} 2\zeta_1\omega_1 & 0 & \dots & 0 \\ 0 & 2\zeta_2\omega_2 & \ddots & \vdots \\ \vdots & \ddots & \ddots & 0 \\ 0 & \dots & 0 & 2\zeta_N\omega_N \end{bmatrix} . \tag{13.45}$$

The covariance matrix $\boldsymbol{\Gamma_{uu}}$ is thus given by

$$\boldsymbol{\Gamma_{uu}}(t) = E\{\boldsymbol{u}(t)\boldsymbol{u}(t)^T\} = E\{\boldsymbol{\Phi z}(t)\boldsymbol{z}(t)^T\boldsymbol{\Phi}^T\} = \boldsymbol{\Phi\Gamma}_{\boldsymbol{yy}}^{\boldsymbol{zz}}(t)\boldsymbol{\Phi}^T . \tag{13.46}$$

In case of filtered white noise excitation, the state and the loading vector, respectively, of the combined system (see (B.1) and (13.2)) in the reduced subspace are given by

$$\boldsymbol{y} = \begin{Bmatrix} \boldsymbol{z} \\ \dot{\boldsymbol{z}} \\ \boldsymbol{v} \end{Bmatrix} , \qquad \boldsymbol{b}(t) = \begin{Bmatrix} \boldsymbol{0} \\ \boldsymbol{0} \\ \boldsymbol{b}_f \end{Bmatrix} , \tag{13.47}$$

where the system matrix for proportional damping is given by

$$\boldsymbol{A} = \begin{bmatrix} \boldsymbol{0} & \boldsymbol{I} & \boldsymbol{0} \\ -\mathrm{diag}(\omega_i^2) & -\mathrm{diag}(2\zeta_i\omega_i) & -\boldsymbol{\Phi}^T\boldsymbol{MI_aQ} \\ \boldsymbol{0} & \boldsymbol{0} & \boldsymbol{A}_f \end{bmatrix} , \tag{13.48}$$

see also (13.44) and (13.45).

13.4 Multi Degree of Freedom System

13.4.1 Model Description

A floor plan of the building is shown in Fig. 13.12. Each of the 12 floors is supported by 72 columns. The floor height is 3.5 m, leading to a total height of 42.0 m. In order to model the dynamic behavior by a multi degree of freedom system with a relatively low number of degrees of freedom, it is assumed that each floor may be represented sufficiently accurate as rigid within the x-y, plane when compared with the flexibility of the columns. Hence, each floor $\{i\}_{i=1}^{12}$ can be represented by just three degrees of freedom, i.e. two displacements u_{xi} and u_{yi} in the x and y direction, respectively, and

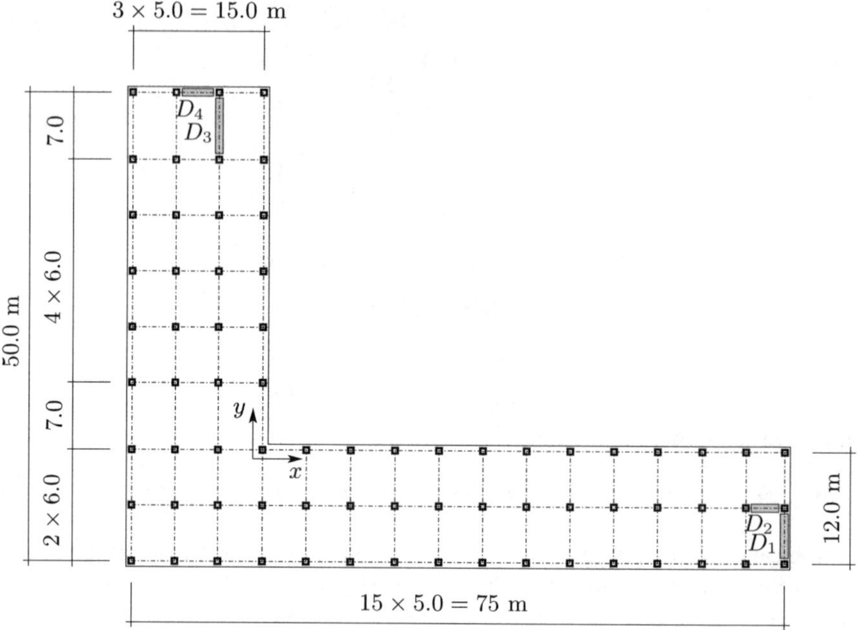

Fig. 13.12. 12-story building: Top view with with hysteretic devices D_1 - D_4.

a rotation u_{zi}. The stiffness between two subsequent floors is assumed to be constant. The sub-matrices ${}^i\boldsymbol{M}$ and ${}^i\boldsymbol{K}$, given by

$$ {}^i\boldsymbol{M} = \begin{bmatrix} 1.3 & 0 & -4.7 \\ 0 & 1.3 & 14.25 \\ -4.7 & 14.25 & 1097 \end{bmatrix} \cdot 10^6 \,, \quad {}^i\boldsymbol{K} = \begin{bmatrix} 1.44 & 0 & -4.88 \\ 0 & 1.44 & 18.0 \\ -4.88 & 18.0 & 1390 \end{bmatrix} \cdot 10^9 \,, \tag{13.49} $$

are referring to the relative displacements and rotation, $u_{r,xi} = u_{x,i} - u_{x,i-1}$, $u_{r,y,i} = u_{y,i} - u_{y,i-1}$ and $u_{r,z,i} = u_{z,i} - u_{z,i-1}$, respectively, using the units kilogram [kg], Newton [N] and meter [m]. The damping sub-matrices ${}^i\boldsymbol{C}$ are assumed to

$$ {}^i\boldsymbol{C} = 0.2388\, {}^i\boldsymbol{M} + 0.002116\, {}^i\boldsymbol{K} \,. \tag{13.50} $$

For an improved earthquake resistance, the structure is reinforced with hysteretic devices as described in Sec. 13.2. In each of the twelve floors, four devices D_1 - D_4 are implemented. The position of these non-linear elements is shown in Fig. 13.12. They provide additional resistance against relative displacements between subsequent floors, where the restoring forces acts for the devices D_1 and D_3 only in the y-direction, and for the devices D_2 and D_4 only in the x-direction. The initial inter-story stiffness, see (13.25), of these devices is identical within each floor, but varies with respect to the floors: Each device has within the floors 1 - 4 the stiffness ${}^I k^{(e)} = 6.0 \cdot 10^8$ N/m, within the floors 5 - 8 the stiffness ${}^{II} k^{(e)} = 4.5 \cdot 10^8$ N/m, and within the floors 9 - 12 the stiffness ${}^{III}k^{(e)} = 3.0 \cdot 10^8$ N/m. These stiffnesses lead to modified inter-story stiffness,

$$ {}^{I}\boldsymbol{K} = \begin{bmatrix} 2.64 & 0 & -24.08 \\ 0 & 2.64 & 51.00 \\ -24.08 & 51.00 & 4453.00 \end{bmatrix} \cdot 10^9 \,, \tag{13.51} $$

$$ {}^{II}\boldsymbol{K} = \begin{bmatrix} 2.34 & 0 & -19.28 \\ 0 & 2.34 & 42.75 \\ -19.28 & 42.75 & 3687.25 \end{bmatrix} \cdot 10^9 \,, \tag{13.52} $$

and

$$ {}^{III}\boldsymbol{K} = \begin{bmatrix} 2.04 & 0 & -14.48 \\ 0 & 2.04 & 34.50 \\ -14.48 & 34.50 & 2921.50 \end{bmatrix} \cdot 10^9 \,, \tag{13.53} $$

which are assembled to the global stiffness matrix $\boldsymbol{K}$ (showing only the upper triangular elements)

$$ \boldsymbol{K} = \begin{bmatrix} 2\,{}^{I}\boldsymbol{K} & -{}^{I}\boldsymbol{K} & \mathbf{0} & \cdots & & & & & \mathbf{0} \\ & 2\,{}^{I}\boldsymbol{K} & -{}^{I}\boldsymbol{K} & \mathbf{0} & \cdots & & & & \mathbf{0} \\ & & 2\,{}^{I}\boldsymbol{K} & -{}^{I}\boldsymbol{K} & \mathbf{0} & \cdots & & & \mathbf{0} \\ & & & {}^{I}\boldsymbol{K} + {}^{II}\boldsymbol{K} & -{}^{II}\boldsymbol{K} & \mathbf{0} & \cdots & & \mathbf{0} \\ & & & & \ddots & \ddots & \ddots & \cdots & \mathbf{0} \\ & & & & & 2\,{}^{III}\boldsymbol{K} & -{}^{III}\boldsymbol{K} & \mathbf{0} & \mathbf{0} \\ & & & & & & 2\,{}^{III}\boldsymbol{K} & -{}^{III}\boldsymbol{K} & \mathbf{0} \\ & & & & & & & 2\,{}^{III}\boldsymbol{K} & -{}^{III}\boldsymbol{K} \\ & & & & & & & & {}^{III}\boldsymbol{K} \end{bmatrix} . \tag{13.54} $$

Additional constants as required in13.30 and 13.31, are: $s_p^{(e)} = 8$ mm for all devices in the floors 1-4, $s_p^{(e)} = 5$ mm for all devices in the floors 5-8, and $s_p^{(e)} = 3$ mm for all devices in the floors 9-12.

13.4.2 Stochastic Excitation

The structure is excited horizontally by earthquake excitation $a_x(t)$ and $a_y(t)$ which is assumed to act independently in the x and y direction, respectively. With respect to (13.1),

$$ \boldsymbol{a}(t) = \begin{Bmatrix} a_x(t) \\ a_y(t) \end{Bmatrix} , \qquad \boldsymbol{Q}_f = \begin{bmatrix} \boldsymbol{Q}_x & \mathbf{0} \\ \mathbf{0} & \boldsymbol{Q}_y \end{bmatrix} , \qquad \boldsymbol{v} = \begin{Bmatrix} \boldsymbol{v}_x \\ \boldsymbol{v}_y \end{Bmatrix} , \tag{13.55} $$

where

$$ \boldsymbol{Q}_x = \boldsymbol{Q}_y = \left\{ \Omega_{1g}^2 \;\; 2\zeta_{1g}\Omega_{1g} \;\; -\Omega_{2g}^2 \;\; -2\zeta_{2g}\Omega_{2g} \right\} . \tag{13.56} $$

The system matrix and the distribution matrix of the filter, respectively, see (13.2), are assumed to

$$ \boldsymbol{A}_f = \begin{bmatrix} \boldsymbol{A}_{fx} & \mathbf{0} \\ \mathbf{0} & \boldsymbol{A}_{fy} \end{bmatrix} , \qquad \boldsymbol{b}_f = \begin{bmatrix} \boldsymbol{b}_{fx} & \mathbf{0} \\ \mathbf{0} & \boldsymbol{b}_{fy} \end{bmatrix} , \tag{13.57} $$

with, see (13.13) and (13.14),

$$\boldsymbol{A}_{fx} = \boldsymbol{A}_{fy} = \begin{bmatrix} 0 & 1 & 0 & 0 \\ -\Omega_{1g}^2 & -2\zeta_{1g}\Omega_{1g} & 0 & 0 \\ 0 & 0 & 0 & 1 \\ \Omega_{1g}^2 & 2\zeta_{1g}\Omega_{1g} & -\Omega_{2g}^2 & -2\zeta_{2g}\Omega_{2g} \end{bmatrix} , \tag{13.58}$$

and

$$\boldsymbol{b}_{fx} = \boldsymbol{b}_{fy} = \begin{Bmatrix} 0 \\ e(t) \\ 0 \\ 0 \end{Bmatrix} . \tag{13.59}$$

For the filter the values $\Omega_{1g} = 15.6$ rad/s, $\zeta_{1g} = 0.6$, $\Omega_{2g} = 1.0$ rad/s and $\zeta_{2g} = 0.995$ have been used. The white noise intensity I has been assumed to be $0.08\ \mathrm{m}^2/\mathrm{s}^3$. For the envelope function $e(t)$ as defined in (13.15) the parameters $a = 0.25$ and $b = 0.5$ have been used. The normalized power spectral density of a_x and a_y corresponding to a stationary ground acceleration, see (13.11), is shown in Fig. 13.13.

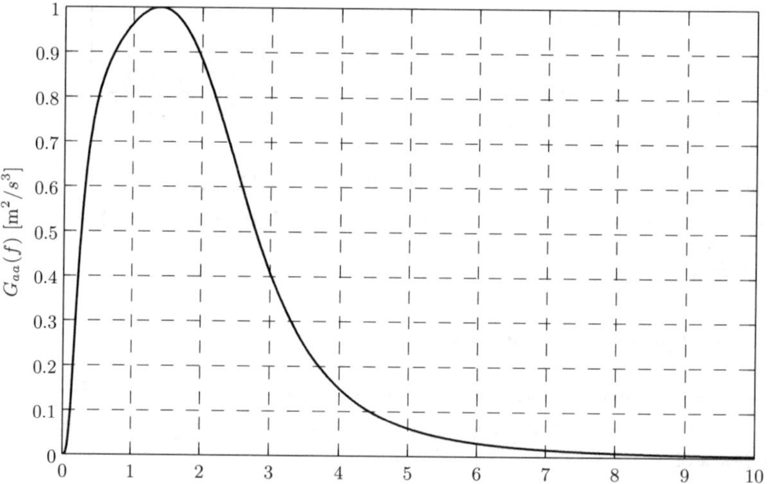

Fig. 13.13. 12-story building: Normalized one-sided auto spectral density $G_{a_xa_x}(f) = G_{a_ya_y}(f)$ for horizontal ground acceleration.

13.4.3 Discussion of Results

In order to demonstrate the accuracy of the proposed procedure, some results are compared with those as presented in [98]. The equivalent modal damping ζ for the reinforced structure has been approximated by

$$\boldsymbol{\phi}_i^T \boldsymbol{C} \boldsymbol{\phi}_j = 2\omega_i \zeta_i \delta_{ij} \; . \tag{13.60}$$

The effect of the non-stationary stochastic response of the reinforced linear structure is computed first, see Fig. 13.14. The damping matrix $\boldsymbol{C}$ is assem-

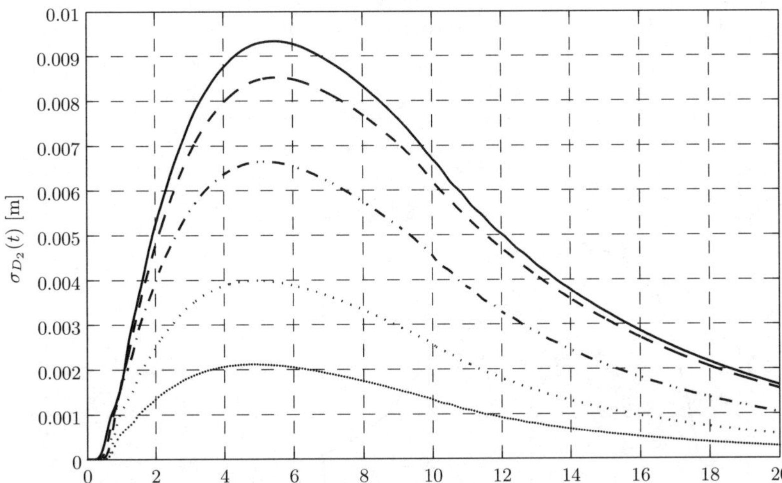

Fig. 13.14. 12-story building: Standard deviation $\sigma_{D_2}(t)$ of relative displacements between floors at the position of device D_2 in x-direction for the linear system obtained by mode acceleration method (10 structural modes) and 100 Karhunen-Loève vectors: 1st floor (solid), 5th floor (dashed), 9th floor (dash-dot), 11th floor (dotted) and 12th floor (points).

bled according to (13.50), and the $\boldsymbol{\phi}$'s are the eigenmodes associated with the reinforced structure. For the structure without friction devices, of course, the exact modal damping corresponding to (13.50) could be calculated according to (6.28).
The convergence behavior regarding the number of Karhunen-Loève vectors and the structural modes, respectively, for the linear structure without devices has been extensively investigated in [122]. Based on these results, $m = 100$ Karhunen-Loève vectors and $N = 10$ structural modes have been used. For the solution based on the Lyapunov matrix differential equation, the excitation has been modeled directly by filtered white noise.

The accuracy of several methods of solution has been investigated: ($\mathcal{A}$) mode acceleration method, see Sec. 6.4, ($\mathcal{B}$) mode displacement method, see Sec. 6.3 in context with 6.4, and ($\mathcal{C}$) direct integration. For time integration the Newmark algorithm, see Sec. 11.2.2, has been applied. A comparison with Fig. 3 in [98] shows excellent agreement between the two different approaches. However, it should be noted that the proposed procedure is much

more straight forward to apply and has the capabilities to incorporate any measured second moment statistics of an excitation process for which the Karhunen-Loève decomposition exists.
For the linear structure, the results obtained by the various methods $(\mathcal{A})$ - $(\mathcal{C})$ basically coincide. The temporal evolution of the standard deviation $\sigma_{D_2}(t)$ of the inter-story displacements for the 5th and 9th floor for the hysteretic system is shown in Fig. 13.15 and 13.16. In order to illustrate the effect of

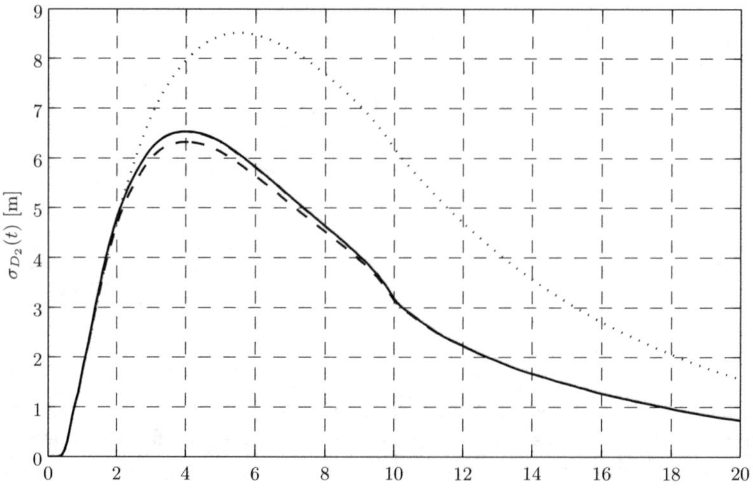

Fig. 13.15. 12-story building: Standard deviation $\sigma_{D_2}(t)$ of relative displacement of 5th floor at the position of device D_2 in x-direction for hysteretic system obtained by 100 Karhunen-Loève vectors applying i) mode acceleration method (10 structural modes, solid); ii) mode displacement method (10 structural modes, dashed); iii) direct integration (dash-dot); iv) response for linear system (dotted).

the hysteretic devices, also the linear response is plotted in Fig. 13.15 and 13.16, see also Fig. 13.14. For the non-linear response it is interesting to note that the approaches $(\mathcal{A})$ - $(\mathcal{C})$ do not yield identical results as it is the case for the linear response. As can be seen, approaches $(\mathcal{A})$ and $(\mathcal{C})$ do agree very well with only minor differences. In this regard it should be mentioned that the modal damping as determined by (13.60) is just an approximation of the proportional damping (13.50) used for direct integration. However, neglecting the response of the higher modes in approach $(\mathcal{B})$ underestimates for instance the maximum value of $\sigma_{D_2}(t)$ about 3% in Fig. 13.15 and overestimates the maximum value of $\sigma_{D_2}(t)$ about 8% in Fig. 13.16.
For comparison: the differences between approaches $(\mathcal{A})$ and $(\mathcal{C})$ plotted in Fig. 13.15 and 13.16 are below 0.4%. The evolution for the mean value of the auxiliary variable $\mu_{q_1}(t) = -\mu q_2(t)$ for different floors is shown in 13.17.

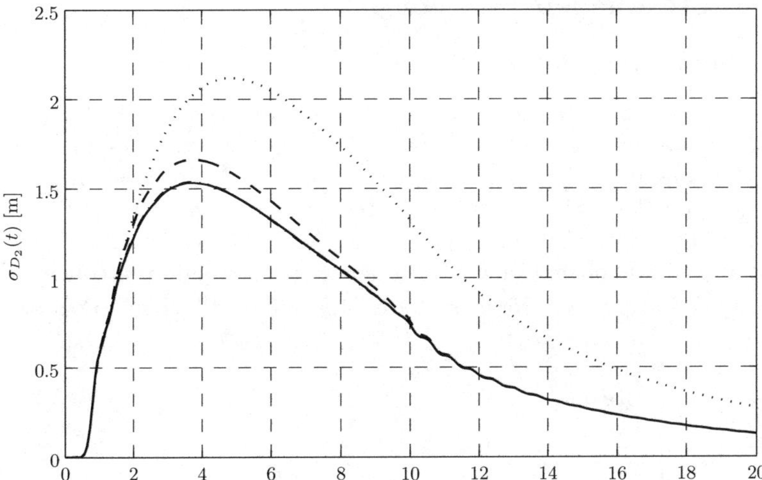

Fig. 13.16. 12-story building: Standard deviation $\sigma_{D_2}(t)$ of relative displacement of 12th floor at the position of device D_2 in x-direction for hysteretic system obtained by 100 Karhunen-Loève vectors applying i) mode acceleration method (10 structural modes, solid); ii) mode displacement method (10 structural modes, dashed); iii) direct integration (dash-dot); iv) response for linear system (dotted).

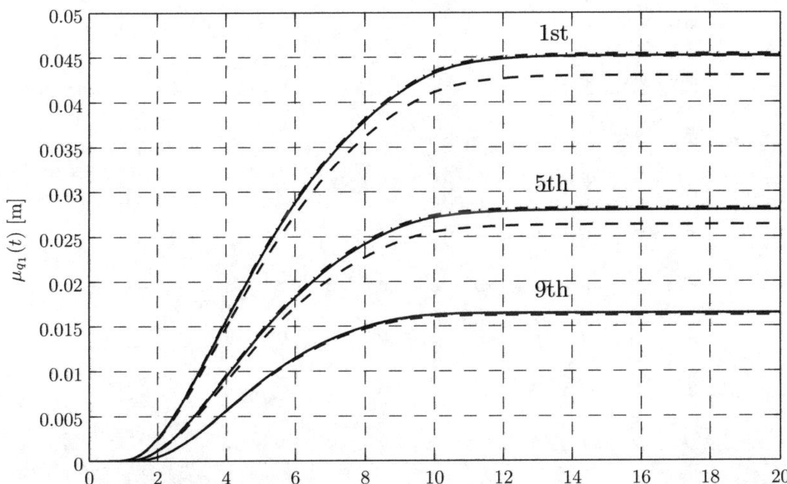

Fig. 13.17. 12-story building: Mean value of auxiliary variable (plastic deformation) $\mu_{q_1}(t)$ of 1st, 5th and 9th floor at the position of device D_2 in x-direction for hysteretic system obtained by 100 Karhunen-Loève vectors applying i) mode acceleration method (10 structural modes, solid); ii) mode displacement method (10 structural modes, dashed); iii) direct integration (dash-dot).

13.5 Finite Element System

13.5.1 Model Description

In order to demonstrate the applicability of the described procedure now for large finite element systems, the stochastic non-stationary response of a 6-story office building with 24,984 degrees of freedom has been analyzed.
The L-shaped floor plan of the reinforced concrete structure is shown in Fig. 13.18, the finite element model consists of 17,351 nodes, 29,592 shell elements and 11,208 beam elements. Properties of the reinforced concrete have been assumed as follows: Young's Modulus $E = 3.0{\cdot}10^{10}$ N/m^2, Poisson ratio

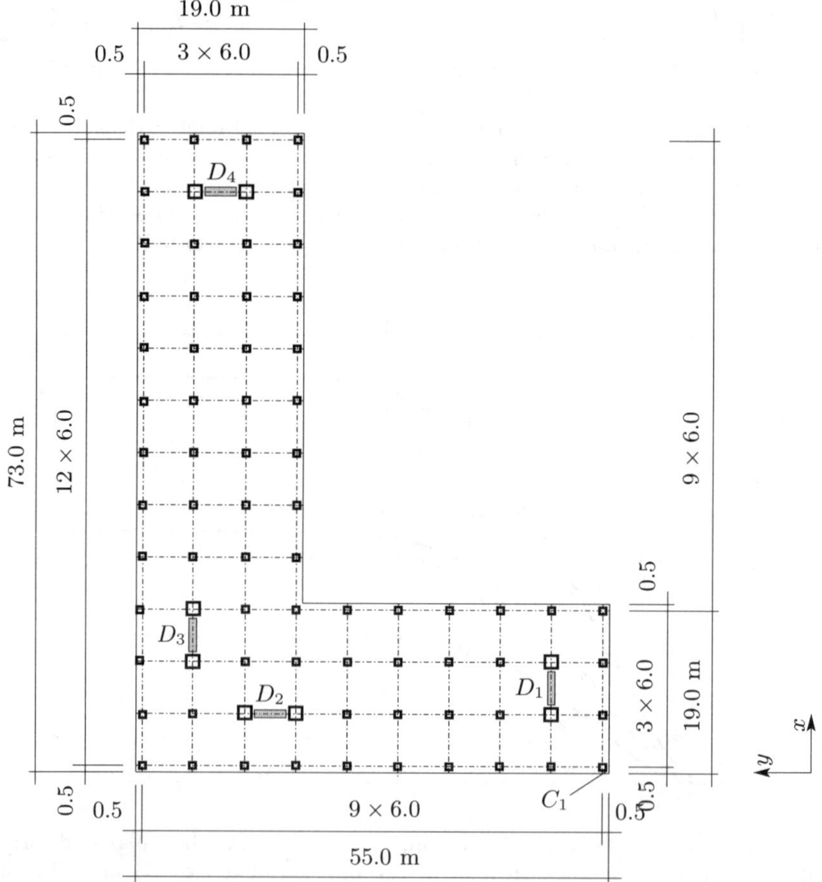

Fig. 13.18. 6-story building; Top view with corner node C_1 and hysteretic devices D_1 - D_4. Dash-dotted lines show the rectangular grid of beam elements used for reinforcement of each floor.

$\mu = 0.3$ and mass density $\varrho = 2500$ kg/m^3. Each floor is modeled by 1249 triangular shell elements with a thickness of 0.25 m involving 3747 degrees of freedom per floor, additionally reinforced by a rectangular grid of beam elements, see dashed-dotted lines in Fig. 13.18. These beams are of rectangular shape with a height of 0.75 m and a width of 0.50 m. The floor height of 3.5 m is constant for all floors, leading to a total height of 84.0 m. Each floor is supported by 76 columns made of reinforced concrete. The outer diameter d of the quadratic cross section of the columns is $d = 0.40$ m, the axial and the bending stiffness is assumed to $EA = 0.48 \cdot 10^{10}$ N, $EI = 0.0064 \cdot 10^{10}$ Nm, respectively. In order to increase the stiffness of the building, reinforced concrete walls from the bottom to the top have been added, located at positions D_1 - D_4 in Fig. 13.18. These walls are again modeled using triangular shell elements with a thickness of 0.25 m and with the same material properties E, μ and ρ as given above.

Non-linear hysteretic shear panels with a material law as described in Sec. 13.2 are located above the reinforced concrete walls and below the upper floor construction, see positions D_1 - D_4 in Fig. 13.18 and Fig. 13.19. The vector

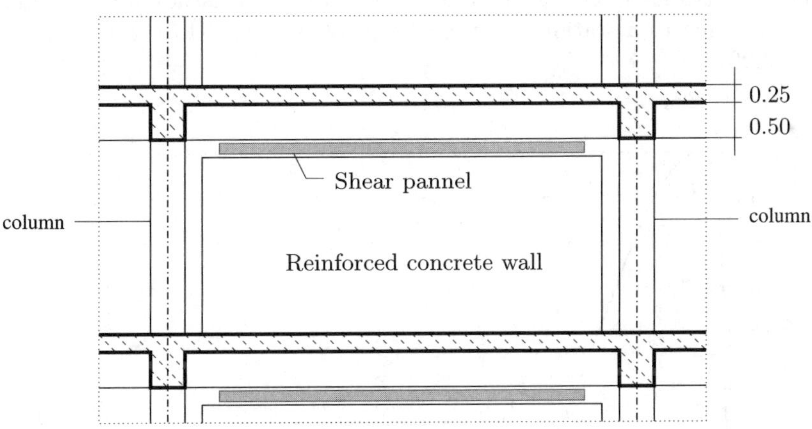

Fig. 13.19. 6-story building: Shear panel.

$\boldsymbol{q}$, see (6.4), comprises thus $6 \times 4 \times 2 = 48$ components.

By building the structural matrices, the initial stiffness of the non-linear elements $k^{(e)}$, see (13.25), has been added to keep the right hand side of the non-linear equation system (6.4) small. The initial inter-story stiffness of these devices is identical within each floor, but varies with respect to the floors: Each device has within the floors 1 - 2 the stiffness $k^{(e)} = 5.0 \cdot 10^8$ N/m, within the floors 3 - 4 the stiffness $k^{(e)} = 4.0 \cdot 10^8$ N/m, and within the floors 5 - 6 the stiffness $k^{(e)} = 3.0 \cdot 10^8$ N/m, the value $s_p^{(e)} = 6$ [mm], see 13.30 and 13.31, has been used for all non-linear devices.

13.5.2 Stochastic Excitation

The structure is excited horizontally and vertically by earthquake excitation $a_x(t)$, $a_y(t)$ and $a_z(t)$, which is assumed to act independently in the x, y and z direction, respectively.
With respect to (13.1), it is assumed that

$$\boldsymbol{Q}_x = \boldsymbol{Q}_y = \boldsymbol{Q}_z = \left\{\Omega_{1g}^2 \; 0 \; -\Omega_{2g}^2 \; -2\zeta_{2g}\Omega_{2g}\right\} , \tag{13.61}$$

the matrices $\boldsymbol{A}_{fx} = \boldsymbol{A}_{fy} = \boldsymbol{A}_{fz}$ are given by (13.58) and the vectors $\boldsymbol{b}_{fx} = \boldsymbol{b}_{fy} = \boldsymbol{b}_{fz}$ are given by (13.59). The values $\Omega_{1g} = 15.0$ rad/s, $\zeta_{1g} = 0.8$, $\Omega_{2g} = 0.3$ rad/s, $\zeta_{2g} = 0.995$, and the white noise intensity $I = 0.18$ m^2/s^3 have been used for the filter associated with the horizontal ground accelerations. For the vertical ground motion, the values $\Omega_{1g} = 20.0$ rad/s, $\zeta_{1g} = 0.8$, $\Omega_{2g} = 1.0$ rad/s, $\zeta_{2g} = 0.995$, and the white noise intensity $I = 0.08$ m^2/s^3 have been used.
For the exponential envelope function given by (13.15), the values $(a, b) = (0.2,0.4)$ for the horizontal motion and $(a, b) = (0.3,0.6)$ for the vertical motion have been assumed. The normalized power spectral densities of a_x, a_y and a_z corresponding to a stationary ground acceleration, see (13.11), are shown in Fig. 13.20.

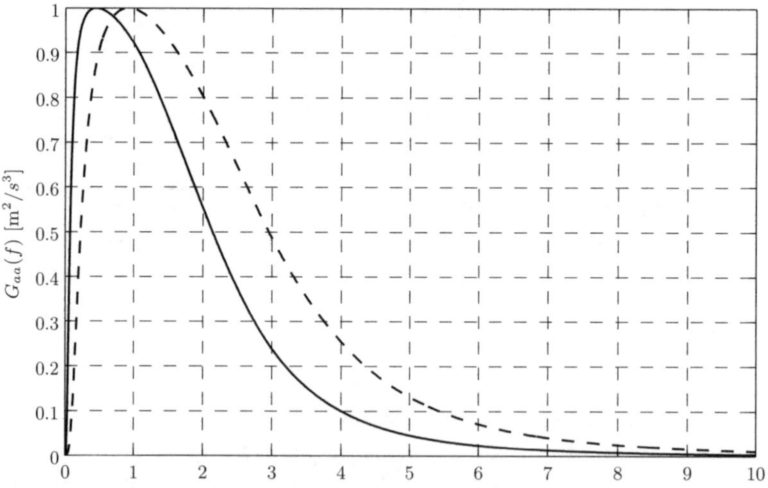

Fig. 13.20. 6-story building: Normalized one-sided auto spectral densities $G_{a_xa_x}(f) = G_{a_ya_y}(f)$ (solid) and $G_{a_za_z}(f)$ (dash-dot) for horizontal and vertical ground acceleration.

13.5.3 Discussion of Numerical Results

Analogous to Sec. 13.4.3, all results have been obtained by using $m = 100$ Karhunen-Loève vectors. Again, several methods of solution have been compared: ($\mathcal{A}$) mode acceleration method using 109 structural modes, ($\mathcal{B}$) mode displacement method using 109 structural modes, ($\mathcal{C}$) direct integration and ($\mathcal{D}$) integration of the modal form of the Lyapunov matrix differential equation using 109 structural modes, see Sec. 13.3). For approaches ($\mathcal{A}$) - ($\mathcal{C}$) the Newmark algorithm and for approach ($\mathcal{D}$) a Runge-Kutta algorithm has been applied, respectively.

The modal damping for all 109 structural modes has been assumed to $\{\zeta\}_{i=1}^{109} = 3\%$ while the proportional damping matrix $\boldsymbol{C}$ for direct integration has been determined by using the relation

$$\boldsymbol{C} = 0.28895\boldsymbol{M} + 0.0021354\boldsymbol{K} \,. \tag{13.62}$$

This is equivalent to $\zeta_1 = \zeta_6 = 3\%$. The standard deviation σ_{C_1} has been computed applying approaches ($\mathcal{A}$) - ($\mathcal{D}$) and is shown in Fig. 13.21 and 13.22. Similar to the 12-story building, the differences between the four approaches are rather small for the linear system, in particular the differences between ($\mathcal{A}$) and ($\mathcal{B}$) are not even visible in Fig. 13.21 - 13.22.

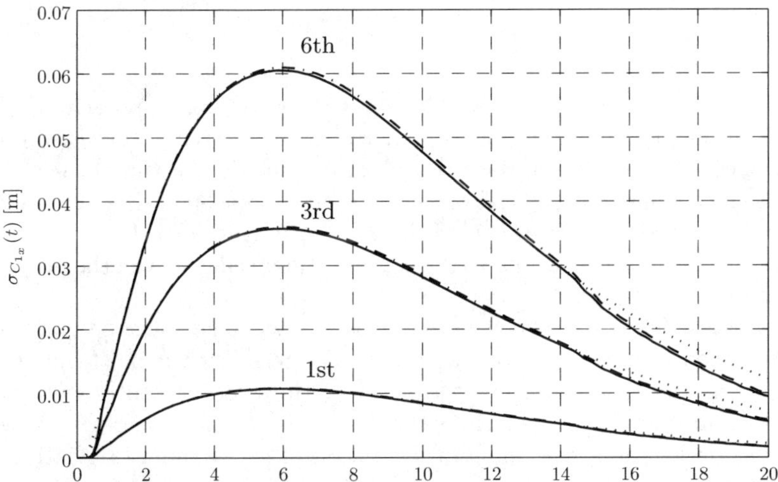

Fig. 13.21. 6-story building: Standard deviation $\sigma_{C_{1_x}}(t)$ of absolute displacement of 1st, 3rd and 6th floor at the position C_1 in x-direction for linear system obtained by 100 Karhunen-Loève vectors applying i) mode acceleration method (109 structural modes, solid); ii) mode displacement method (109 structural modes, dashed); iii) direct integration (dash-dot); iv) integration of modal form of Lyapunov matrix differential equation (dotted).

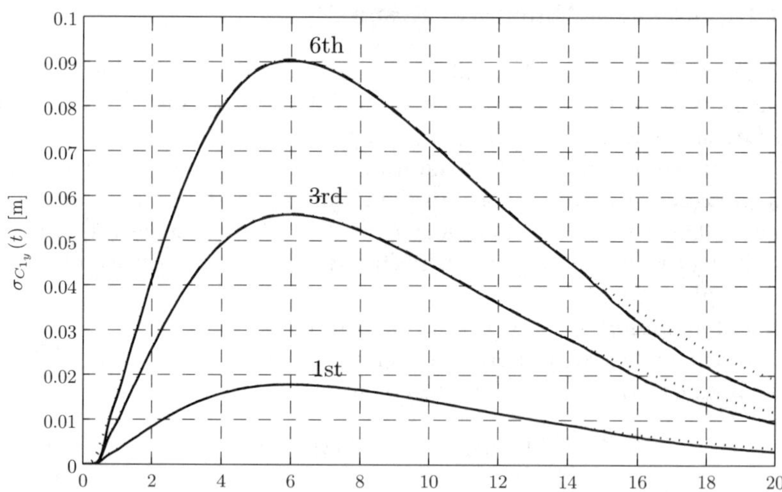

Fig. 13.22. 6-story building: Standard deviation $\sigma_{C_{1_y}}(t)$ of absolute displacement of 1st, 3rd and 6th floor at the position C_1 in y-direction for linear system obtained by 100 Karhunen-Loève vectors applying i) mode acceleration method (109 structural modes, solid); ii) mode displacement method (109 structural modes, dashed); iii) direct integration (dash-dot); iv) integration of modal form of Lyapunov matrix differential equation (dotted).

Some remarks on Fig. 13.21: As already mentioned, for approach $(\mathcal{D})$ the excitation has been modeled by filtered white noise. This is the reason why approaches $(\mathcal{A})$ - $(\mathcal{C})$ do deviate considerably outside the time range [1,14] seconds from approach $(\mathcal{D})$. The accuracy of approaches $(\mathcal{A})$ - $(\mathcal{C})$ outside this time range could be increased, if the number of Karhunen-Loève vectors is increased, which in turn decreases the computational efficiency of the proposed procedure.

Another possible way of increasing the accuracy while keeping the number of the Karhunen-Loève vectors constant would be to select the most important Karhunen-Loève vectors for every time step. However, as can be seen from Fig. 13.21, in the time range [1,14] seconds, all approaches do agree very well. At the same location, the temporal evolution of the standard deviation for the hysteretic system has been computed and plotted for various floors in Fig. 13.23 and Fig. 13.24. Again, also the linear response is shown in order to illustrate the effect of the hysteretic devices. As it has been observed for the 12-story building, the static response of the higher modes contributes significantly to the response of the non-linear system.

The time needed for integrating the hysteretic system in the time range [0,10] seconds was about 5.5 hours and 120 hours for approach $(\mathcal{A})$ and $(\mathcal{C})$ on a PC with 1,400 MHz, respectively, which demonstrates the computational

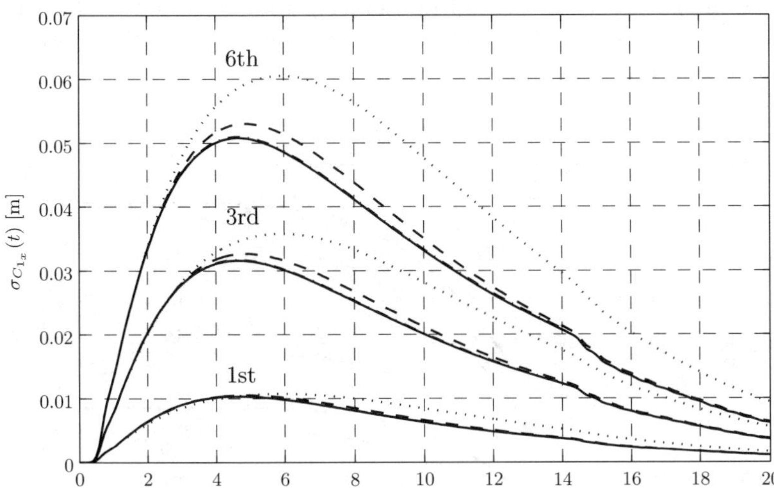

Fig. 13.23. 6-story building: Standard deviation $\sigma_{C_{1x}}(t)$ of absolute displacement of 1st, 3rd and 6th floor at the position C_1 in x-direction for hysteretic system obtained by 100 Karhunen-Loève vectors applying i) mode acceleration method (109 structural modes, solid); ii) mode displacement method (109 structural modes, dashed); iii) direct integration (dash-dot); iv) response for linear system (dotted).

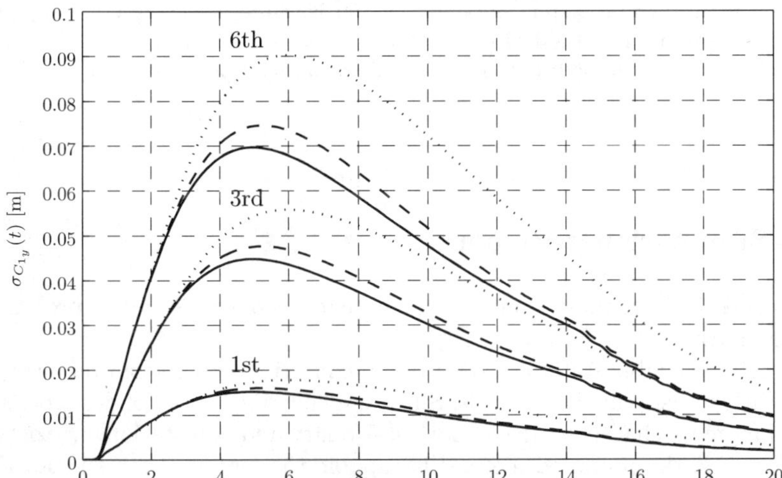

Fig. 13.24. 6-story building: Standard deviation $\sigma_{C_{1y}}(t)$ of absolute displacement of 1st, 3rd and 6th floor floor at the position C_1 in x-direction for hysteretic system obtained by 100 Karhunen-Loève vectors applying i) mode acceleration method (109 structural modes, solid); ii) mode displacement method (109 structural modes, dashed); iii) direct integration (dash-dot); iv) response for linear system (dotted).

efficiency of the proposed method by reducing the dimension of the system equations using modal analysis. The mean function of the plastic deformation μ_{q_1} is shown in Fig. 13.25.

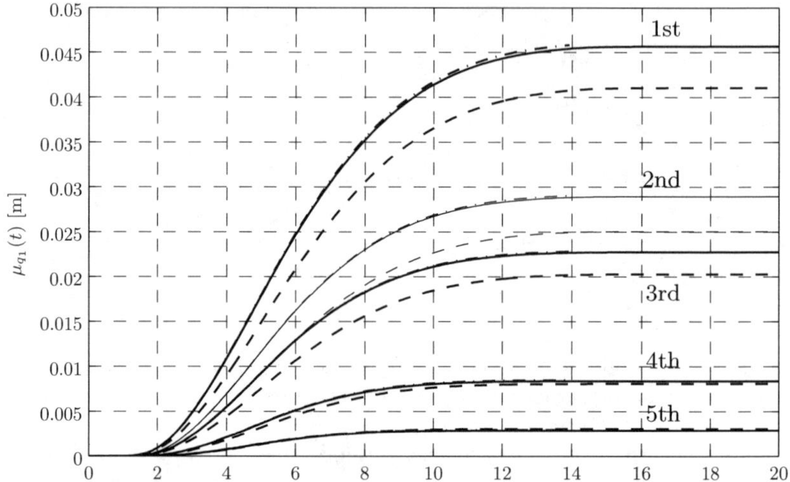

Fig. 13.25. 6-story building: Mean value of auxiliary variable (plastic deformation) $\mu_{q_1}(t)(t)$ of 1st, 2nd, 3rd, 4th and 5th floor at the position of device D_1 in x-direction for hysteretic system obtained by 100 Karhunen-Loève vectors applying i) mode acceleration method (10 structural modes, solid); ii) mode displacement method (10 structural modes, dashed); iii) direct integration (dash-dot).

13.6 Summarizing Remarks

At the end of this chapter, some important features of the described procedure are summarized.

Potentially available statistical data of an excitation process can be incorporated rather easily in the analysis. Since the covariance matrix describes the second moment characteristics of both stationary and non-stationary stochastic processes, the proposed method is capable to compute both the stationary and the non-stationary system response, respectively, i.e. in an efficient manner. Many of the simplifications and physical inconsistencies, as they are introduced by procedures relying on white noise models of the excitation process, are avoided by the present approach. This applies, for instance, to the well known problem that the white noise model introduces unrealistic high frequency excitations and thus unrealistic responses. Contrary to many other approaches in stochastic dynamics, non-zero mean problems can be handled

very efficiently. The proposed approach relies on deterministic integration algorithms, thus limitations on the size of the system are comparable to those encountered in a purely deterministic analysis.
As a minor limitation, the quality of the approximation of the actual system response depends clearly on the quality or accuracy of the equivalent statistical linearization technique. This also holds for the level of non-linearity that can be dealt with. The use of equivalent statistical linearization requires additional codes when applying commercial finite element packages, since so far equivalent statistical linearization is not yet part of these packages.

Part IV

Appendices

A

Imperfection Data Bank

For the stochastic modeling of geometric imperfections as described in Sec. 12.1, a database [5] has been available. This database provides comprehensive information to be used to improve the quality of the numerical prediction of the stability behavior of cylindrical shells. It includes measured imperfections of cylindrical shells of different sizes manufactured by different processes as well as experimentally determined buckling loads. In this work, seven copper electro-plated shells, referred in [5] as A-Shells, are statistically analyzed in order to model the geometric imperfections most realistically, see Table A.1. These laboratory scaled cylindrical shells are about 200 mm in length and the radius is about 100 mm. The largest imperfection magnitude referenced to the so-called perfect cylinder is in the range of 0.47 mm, which is approximately three times the shell thickness. [5] showed, that the axi-symmetric imperfections (parallel to the shell axis) are in general about an order smaller than the axi-symmetric imperfections. In Figs. A.1–A.4 the geometric imperfections are unwound from the perfect cylinder. It should be noted, that these plots should provide a good visual control of the imperfections, i.e. they are not to scale. To estimate the covariance matrix of the geometric imperfections according to (12.3), the origin of the coordinate system has to be defined. For this purpose, the position of the largest imperfection magnitude of every shell is chosen to define the origin of the shell in circumferential direction. For a grouping of the shells in axial direction, a notably varying mean-square value along the axial direction was utilized.
Second moment characteristics of boundary imperfections, see Sec. 12.1, have been determined from one sample shown in Fig. A.5, see [4].

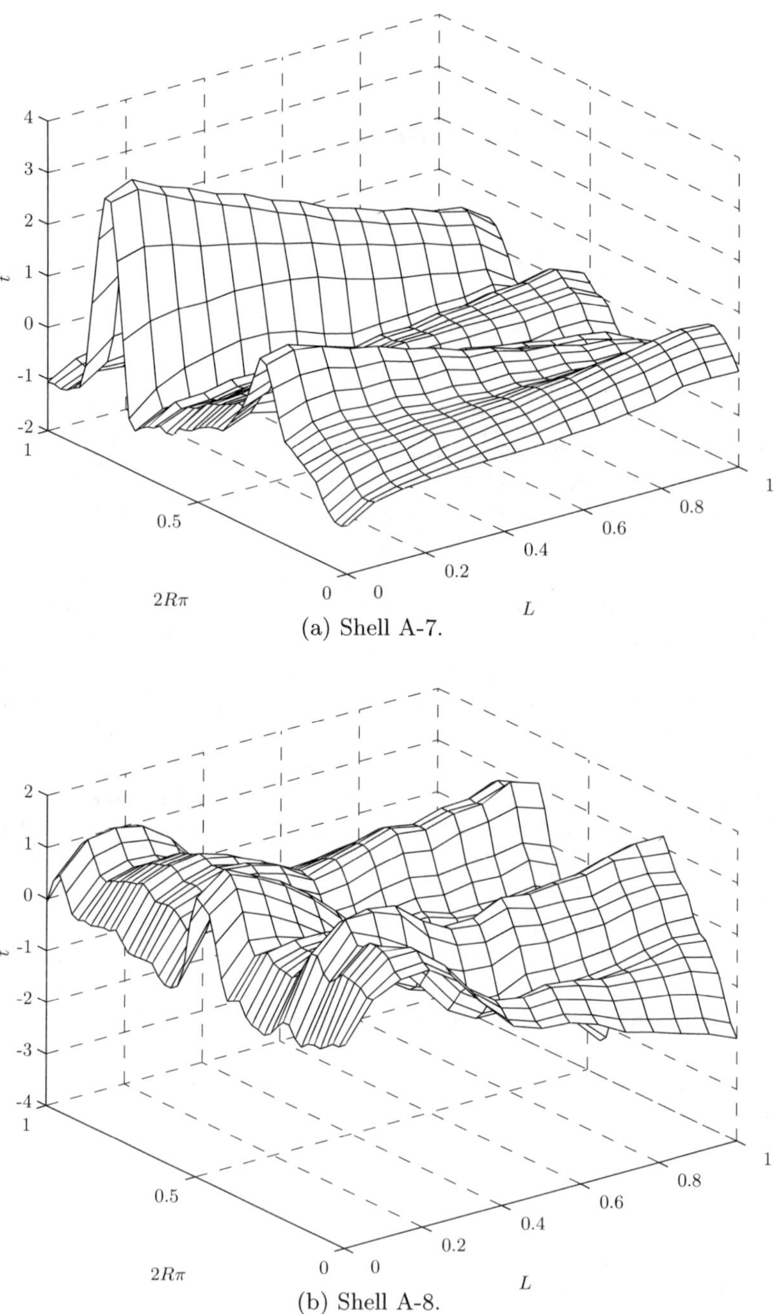

(a) Shell A-7.

(b) Shell A-8.

Fig. A.1. Geometric imperfections of electro-plated isotropic cylindrical shells (A7 and A8).

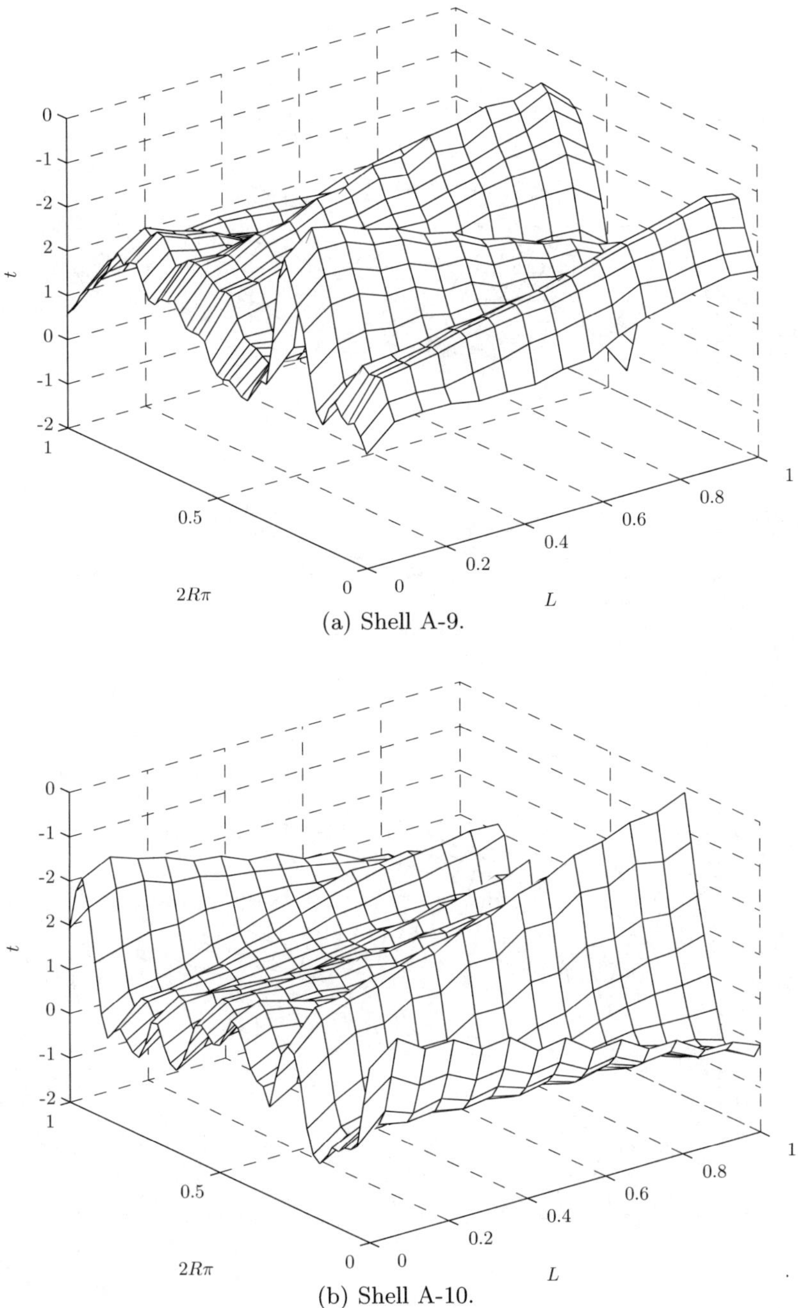

(a) Shell A-9.

(b) Shell A-10.

Fig. A.2. Geometric imperfections of electro-plated isotropic cylindrical shells (A9 and A10).

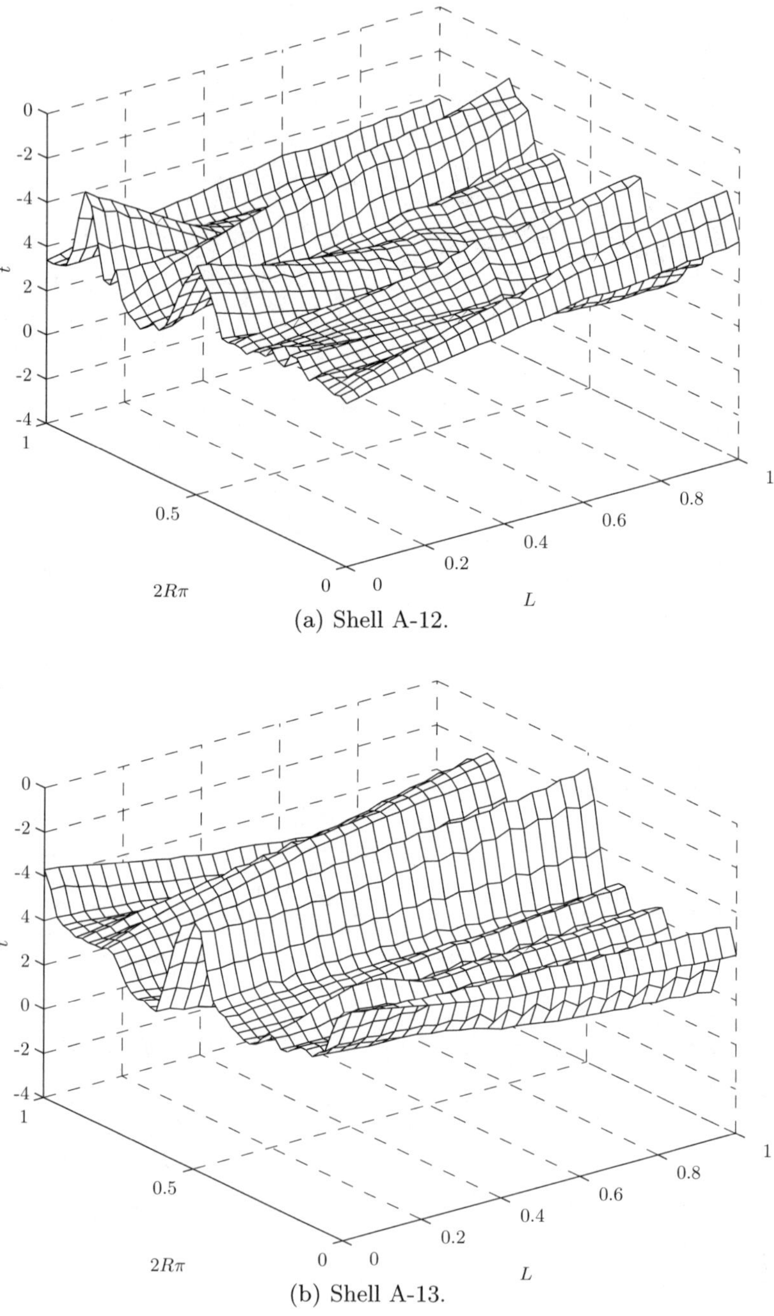

(a) Shell A-12.

(b) Shell A-13.

Fig. A.3. Geometric imperfections of electro-plated isotropic cylindrical shells (A12 and A13).

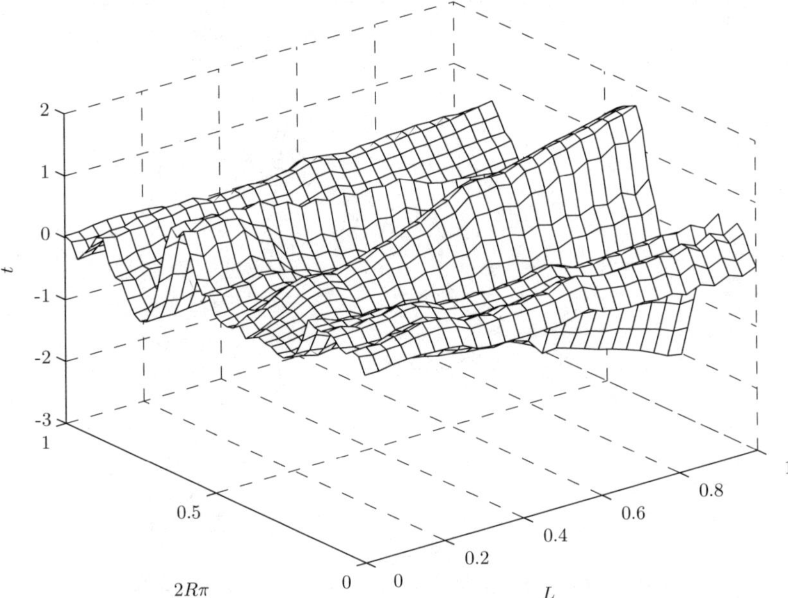

Fig. A.4. Geometric imperfections of electro-plated shell A-14.

Shell	R [m]	t [m]	L [m]	E [N/m^2]	P [N]	$\lambda_{\mathrm{exp}}^{\mathrm{lim}}$
A-7	0.1016	0.000114	0.20320	$1.0411 \cdot 10^{11}$	5145.18	0.5901
A-8	0.1016	0.000118	0.20320	$1.0480 \cdot 10^{11}$	5539.72	0.6632
A-9	0.1016	0.000115	0.20320	$1.0135 \cdot 10^{11}$	5123.67	0.7270
A-10	0.1016	0.000120	0.20320	$1.0273 \cdot 10^{11}$	5663.03	0.5645
A-12	0.1016	0.000120	0.20955	$1.0480 \cdot 10^{11}$	5777.14	0.6669
A-13	0.1016	0.000113	0.19685	$1.0411 \cdot 10^{11}$	5037.43	0.6171
A-14	0.1016	0.000111	0.19685	$1.0894 \cdot 10^{11}$	5104.25	0.6745

Table A.1. Dimensions, material properties and experimentally determined limit loads, respectively, of the seven A-shells [5]. Radius R, thickness t, length L, Young's modulus E, measured limit load P and corresponding normalized load level $\lambda_{\mathrm{exp}}^{\mathrm{lim}}$.

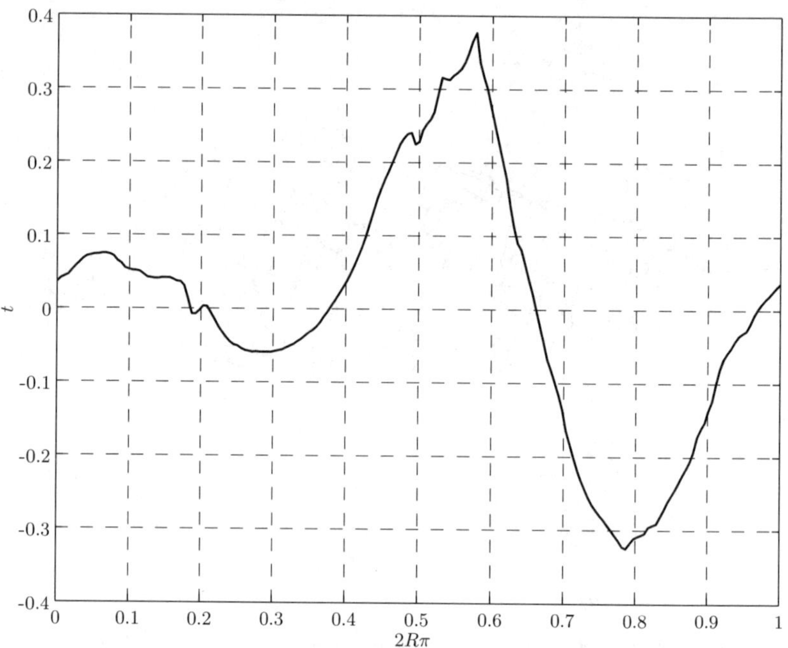

Fig. A.5. Measured boundary imperfections [4].

B

Lyapunov Matrix Differential Equation

The state space equation of the system defined in (6.18) is given by, see e.g. [91],

$$\dot{\boldsymbol{y}}(t) = \boldsymbol{A}\boldsymbol{y}(t) + \boldsymbol{b}(t) \; , \tag{B.1}$$

where the state vector and the force vector are defined by

$$\boldsymbol{y}(t) = \begin{Bmatrix} \boldsymbol{u} \\ \dot{\boldsymbol{u}} \end{Bmatrix} , \qquad \boldsymbol{b}(t) = \begin{Bmatrix} \boldsymbol{0} \\ \boldsymbol{M}^{-1}\boldsymbol{f}(t) \end{Bmatrix} , \tag{B.2}$$

respectively. Using the symbol $\boldsymbol{I}$ for the identity matrix, the system matrix is given by

$$\boldsymbol{A} = \begin{bmatrix} \boldsymbol{0} & \boldsymbol{I} \\ -\boldsymbol{M}^{-1}\boldsymbol{K} & -\boldsymbol{M}^{-1}\boldsymbol{C} \end{bmatrix} . \tag{B.3}$$

In the following it is assumed, that the loading vector $\boldsymbol{b}(t)$ of (B.1) can be written such that

$$\boldsymbol{b}(t) = \boldsymbol{G}(t)\boldsymbol{c}(t) \tag{B.4}$$

holds, where $\boldsymbol{G}(t)$ is the so-called distribution matrix and $\boldsymbol{c}(t)$ is a the actual loading vector. Commonly, $\boldsymbol{G}(t)$ is time invariant, i.e. $\boldsymbol{G}(t) = \boldsymbol{G}$. The general solution of the first order system (B.1) is then given by, see e.g. [141],

$$\boldsymbol{y}(t) = \boldsymbol{\Upsilon}(t,0)\boldsymbol{y}(0) + \int_0^t \boldsymbol{\Upsilon}(t,\tau)\boldsymbol{G}\,\boldsymbol{c}(\tau)\,\mathrm{d}\tau \; , \tag{B.5}$$

where $\boldsymbol{\Upsilon}(t,\tau)$ is the so-called principal or fundamental matrix and $\boldsymbol{y}(0)$ denote the stochastic initial conditions. $\boldsymbol{\Upsilon}(t,0)$ can be given for time invariant systems in the explicit form of

$$\boldsymbol{\Upsilon}(t,0) = \mathrm{e}^{\boldsymbol{A}t} = \sum_{j=0}^{\infty} \frac{\boldsymbol{A}^j t^j}{j!} . \tag{B.6}$$

The principal matrix for different time steps can be obtained by the linear relation

$$\boldsymbol{\Upsilon}(t_2, 0) = \boldsymbol{\Upsilon}(t_2, t_1)\boldsymbol{\Upsilon}(t_1, 0) \ . \tag{B.7}$$

In order to obtain the covariance matrix of the response

$$\boldsymbol{\Gamma}_{\boldsymbol{yy}}(t,s) = E\{(\boldsymbol{y}(t) - \boldsymbol{\mu}_{\boldsymbol{y}}(t))(\boldsymbol{y}(s) - \boldsymbol{\mu}_{\boldsymbol{y}}(s))^T\} \ , \tag{B.8}$$

the centered (zero mean) response and forcing processes, respectively, are needed, i.e.

$$\dot{\tilde{\boldsymbol{y}}}(t) = \dot{\boldsymbol{y}}(t) - \boldsymbol{\mu}_{\boldsymbol{y}}(t) \ , \tag{B.9}$$

and

$$\dot{\tilde{\boldsymbol{c}}}(t) = \dot{\boldsymbol{b}}(t) - \boldsymbol{\mu}_{\boldsymbol{c}}(t) \ . \tag{B.10}$$

Equation (B.1) and (B.5) can then be rewritten according to

$$\dot{\tilde{\boldsymbol{y}}}(t) = \boldsymbol{A}\tilde{\boldsymbol{y}}(t) + \boldsymbol{G}\,\tilde{\boldsymbol{c}}(t) \ , \tag{B.11}$$

and

$$\tilde{\boldsymbol{y}}(t) = \boldsymbol{\Upsilon}(t,0)\tilde{\boldsymbol{y}}(0) + \int_0^t \boldsymbol{\Upsilon}(t,\tau)\boldsymbol{G}\,\tilde{\boldsymbol{c}}(\tau)\,d\tau \ . \tag{B.12}$$

Post-multiplying (B.11) with $\tilde{\boldsymbol{y}}^T(s)$ and substituting (B.12) yields

$$\dot{\tilde{\boldsymbol{y}}}(t)\tilde{\boldsymbol{y}}^T(s) = \boldsymbol{A}\tilde{\boldsymbol{y}}(t)\tilde{\boldsymbol{y}}^T(s) + \boldsymbol{G}\,\tilde{\boldsymbol{c}}(t)\Big[\boldsymbol{\Upsilon}(s,0)\tilde{\boldsymbol{y}}(0) + \int_0^t \boldsymbol{\Upsilon}(s,\tau)\boldsymbol{G}\,\tilde{\boldsymbol{c}}(\tau)\,d\tau\Big]^T \ , \tag{B.13}$$

which one allows to derive a differential equation for the covariance matrix between different time instants by taking the expectation of (B.13)

$$\frac{\partial}{\partial t}\boldsymbol{\Gamma}_{\boldsymbol{yy}}(t,s) = \boldsymbol{A}\boldsymbol{\Gamma}_{\boldsymbol{yy}}(t,s) + \boldsymbol{D}(t,s) \ . \tag{B.14}$$

It should be noted that (B.14) has been obtained under the assumption that the loading and the initial conditions are uncorrelated. The matrix $\boldsymbol{D}(t,s)$, commonly denoted as the spectral matrix, is defined as

$$\boldsymbol{D}(t,s) = \boldsymbol{G}\int_0^s \boldsymbol{\Gamma}_{\boldsymbol{cc}}(t,\tau)\boldsymbol{G}^T\boldsymbol{\Upsilon}^T(s,\tau)\,d\tau \ . \tag{B.15}$$

The so-called Lyapunov matrix differential equation describes the temporal evolution of the covariances and can be derived using the relation

$$\dot{\boldsymbol{\Gamma}}_{\boldsymbol{yy}}(t,t) = \dot{\boldsymbol{\Gamma}}_{\boldsymbol{yy}}(t) = E\Big\{\frac{d}{dt}(\tilde{\boldsymbol{y}}(t)\tilde{\boldsymbol{y}}^T(t))\Big\} = E\Big\{\dot{\tilde{\boldsymbol{y}}}(t)\tilde{\boldsymbol{y}}^T(t) + \tilde{\boldsymbol{y}}(t)\dot{\tilde{\boldsymbol{y}}}^T(t)\Big\} \ , \tag{B.16}$$

(B.10) and (B.11), yielding

$$\dot{\boldsymbol{\Gamma}}_{\boldsymbol{yy}}(t) = \boldsymbol{A}\boldsymbol{\Gamma}_{\boldsymbol{yy}}(t) + \boldsymbol{\Gamma}_{\boldsymbol{yy}}(t)\boldsymbol{A}^T + \boldsymbol{D}(t) + \boldsymbol{D}^T(t) \ . \tag{B.17}$$

Equation (B.17) is rather difficult to solve because of the integration of (B.15). However, for white noise excitation $\boldsymbol{w}(t)$, i.e. $\boldsymbol{c}(t) = \boldsymbol{w}(t)$ and thus

$$\boldsymbol{\Gamma}_{\boldsymbol{ww}}(t,\tau) = E\{\boldsymbol{w}(t)\boldsymbol{w}^T(t+\tau)\} = \boldsymbol{I}_w(t)\delta(\tau)\ , \tag{B.18}$$

where $\delta(\tau)$ and $\boldsymbol{I}_{\boldsymbol{w}}(t)$ denote the Dirac's delta function and the white noise intensity, respectively, the spectral matrix

$$\boldsymbol{D}(t,s) = \boldsymbol{G}\int_0^s \boldsymbol{I}_{\boldsymbol{w}}(\tau)\delta(t-\tau)\boldsymbol{G}^T\boldsymbol{\Upsilon}^T(s,\tau)\,\mathrm{d}\tau\ , \tag{B.19}$$

is zero for $t > s$. The white noise intensity is related to the power spectral density $\boldsymbol{S}_0(t)$ of the white noise loading by

$$\boldsymbol{I}_{\boldsymbol{w}}(t) = 2\pi\boldsymbol{S}_0(t)\ . \tag{B.20}$$

Keeping in mind that $\boldsymbol{\Upsilon}(t,t) = \boldsymbol{I}$, where $\boldsymbol{I}$ denotes the identity matrix, and that a white noise impulse is responsible for a jump in the velocity response for the time interval $t-\epsilon \le t \le t+\epsilon$, where ϵ is an arbitrary small number, one arrives for $t = s$, see e.g. [72], at

$$\boldsymbol{D}(t,t) = \frac{1}{2}\boldsymbol{G}\,\boldsymbol{I}_{\boldsymbol{w}}(t)\boldsymbol{G}^T\ . \tag{B.21}$$

The Lyapunov matrix differential equation for white noise loading is thus given by

$$\dot{\boldsymbol{\Gamma}}_{\boldsymbol{yy}}(t) = \boldsymbol{A}\boldsymbol{\Gamma}_{\boldsymbol{yy}}(t) + \boldsymbol{\Gamma}_{\boldsymbol{yy}}(t)\boldsymbol{A}^T + \boldsymbol{G}\boldsymbol{I}_{\boldsymbol{w}}(t)\,\boldsymbol{G}^T\ . \tag{B.22}$$

(B.22) is also valid for colored noise loading, as shown in Chap. 13.1.

References

1. B.O. Almroth and M.C. Holmes. Buckling of shells with cutouts, experiment and analysis. *International Journal of Solids and Structures*, 8:1057–1071, 1972.
2. J. C. Amazigo. Buckling under axial compression of long cylindrical shells with random axisymmetric imperfections. *Quarterly Journal of Applied Mathematics*, 26(4):537–566, 1969.
3. D.R. Ambur, N. Jaunky, M.W. Hilburger, and C. G. Davila. Progressive failure analyses of compression-loaded composite curved panels with and without cutouts. *Composite Structures*, 65(2):143–155, 2004.
4. J. Arbocz. The effect of imperfect boundary conditions on the collapse behavior of anisotropic shells. *International Journal of Solids and Structures*, 37:6891–6915, 2000.
5. J. Arbocz and H. Abramovich. The Initial Imperfection Data Bank at the Delft University of Technology, Part 1. Technical Report LR-290, Delft University of Technology, Department of Aerospace Engineering, December 1979.
6. J. Arbocz and J. M. A. M. Hol. Koiter's stability theory in a computer aided engineering (CAE) environment. *International Journal of Solids and Structures*, 26(9-10):945–973, 1991.
7. J. Arbocz and J.M.A.M. Hol. Collapse of axially compressed cylindrical shells with random imperfections. *Thin-Walled Structures*, 23(1-4):131–158, 1995.
8. J. Argyris, M. Papadrakakis, and G. Stefanou. Stochastic finite element analysis of shells. *Computer Methods in Applied Mechanics and Engineering*, 191(41-42):4781–4804, 2002.
9. T.S Atalik and S. Utku. Stochastic linearization of multi-degree-of-freedom non-linear systems. *Earthquake Engineering and Structural Dynamics*, 4:411–420, 1976.
10. S.K. Au and J.L. Beck. First excursion probabilities for linear systems by very efficient importance sampling. *Probabilistic Engineering Mechanics*, (16):193–207, 2001.
11. M. F. Azeez and A. F. Vakakis. Proper orthogonal decomposition (POD) of a class of vibroimpact oscillations. *Journal of Sound and Vibration*, 240(5):859–889, 2001.
12. T.T. Baber and Y.K. Wen. Random vibration of hysteretic degrading systems. *Journal of Engineering Mechanics*, 107(6):1069–1086, 1981.

13. Y. Basar and W. B. Krätzig. *Mechanik der Flächentragwerke, Theorie, Berechnungsmethoden, Anwendungsbeispiele.* Friedr. Vieweg & Sohn, Braunschweig, Wiesbaden, 1985.
14. K. J. Bathe. *Finite element procedures.* Prentice Hall, Upper Saddle River, New Jersey, 1996.
15. Z. P. Bazant and L. Cedolin. *Stability of structures.* Oxford University Press, New York, Oxford, 1 edition, 1991.
16. T. Belytschko, W. K. Liu, and B. Moran. *Nonlinear Finite Elements for Continua and Structures.* John Wiley & Sons, Ltd, Chichester, New York, 2000.
17. E. Bielewicz and J. Górski. Shells with random geometric imperfections simulation - based approach. *International Journal of Non-Linear Mechanics*, 37(4-5):777–784, 2002.
18. E. Bielewicz, J. Górski, R. Schmidt, and H. Walukiewicz. Random fields in the limit analysis of elastic-plastic shell structures. *Computers and Structures*, 51(3):267–275, 1994.
19. V. V. Bolotin. Statistical methods in the nonlinear theory of elastic shells. Izvestiya AN SSSR, Otdeleniye Teknicheskikh Nauk, N 3. Technical report, 1958. English translation: NASA TTF-85, 1-16, 1962.
20. H. Bremer. *Dynamik und Regelung mechanischer Systeme.* Teubner Studienbücher Mechanik, Stuttgart, 1988. Leitfäden der angewandten Mathematik und Mechanik; Bd. 67.
21. C. E. Brenner. Stochastic finite element methods (literature review). Technical Report 35-91, Institute of Engineering Mechanics, Leopold-Franzens University, Innsbruck, Austria, EU, November 1991.
22. F. A. Brogan, C. C. Rankin, and H. D. Cabiness. *STAR Reference Manual, LMMS P032595.* Lockheed Martin Missiles and Space Co., Inc., Palo Alto, CA, USA, 1996.
23. Barry W. Brown, James Lovato, Kathy Russel, and John Venier. *RANDLIB Library of Routines for Random Number Generation (C and Fortran).* http://utmdacc.mdacc.tmc.edu/.
24. D. Bushnell. Computerized analyisis of shells - governing equations. *Computers and Structures*, 18(3):471–536, 1984.
25. D. Bushnell. *Computerized Analysis of Shells (Mechanics of Elastic Stability).* Kluwer Academic Publishers, Dordrecht, Boston, London, 1989.
26. G.Q. Cai and Y.K. Lin. Generation of non-Gaussian stationary processes. *Physical Review E*, 54(1):299–303, 1996.
27. R. Cameron and W. Martin. The orthogonal development of nonlinear functionals in series of Fourier-Hermite functionals. *Ann. Math*, 48:385–392, 1947.
28. G. Cederbaum and J. Arbocz. On the reliability of imperfection-sensitive long isotropic cylindrical shells. *Structural Safety*, 18(1):1–9, 1996.
29. T.P. Chang, T. Mochio, and E. Samaras. Seismic response analysis of nonlinear structures. *Probabilistic Engineering Mechanics*, 1(3):157–166, 1986.
30. D. Chapelle and K.J. Bathe. *The Finite Element Analysis of Shells.* Computational Fluid and Solid Mechanics. Springer, Berlin, Heidelberg, New York, 2003.
31. M.K. Chryssanthopoulos and C. Poggi. Probabilistic imperfection sensitivity analysis of axially compressed composite cylinders. *Engineering Structures*, 17(6):398–406, 1995.
32. M.K. Chryssanthopoulos and C. Poggi. Stochastic imperfection modelling in shell buckling studies. *Thin-Walled Structures*, 23(1-4):179–200, 1995.

33. R.W. Clough and J. Penzien. *Dynamics of Structures*. McGraw-Hill International Editions, 2nd edition, 1993.
34. R. Courant and D. Hilbert. *Methoden der mathematischen Physik*. Springer Verlag, 3 edition, 1967.
35. R.R. Jr. Craig. *Structural Dynamics, An introduction to computer methods*. John Wiley and Sons, New York, Chichester, Brisbane, Toronto, 1981.
36. S. H. Crandall. *Random Vibration in Mechanical Systems*. Academic Press Inc., 1963.
37. M. A. Crisfield. *Non-linear Finite Element Analysis of Solids and Structures*, volume 1: Essentials. John Wiley & Sons, Ltd, Chichester, New York, 1991.
38. I. Daubechies. *Ten Lectures on Wavelets*. John Wiley and Sons, New York, 1992.
39. W. B. Davenport and L. R. Williams. *An Introduction to the Theory of Random Signals and Noise*. McGraw-Hill Book Company, 1958.
40. G. Deodatis and M. Shinozuka. Simulation of seismic ground motion using stochastic waves. *Journal of Engineering Mechanics*, 115(12):2723–2737, 1989.
41. I. Elishakoff and J. Arbocz. Reliability of axially compressed shells with random axisymmetric imperfections. *International Journal of Solids and Structures*, 18(7):563–585, 1982.
42. I. Elishakoff and J. Arbocz. Reliability of axially compressed cylindrical shells with general nonsymmetric imperfections. *Journal of Applied Mechanics*, 52:122–128, 1985.
43. H.H. Emam, H.P. Pradlwarter, and G.I. Schuëller. A computational procedure for the implementation of equivalent linearization in finite element analysis. *Journal of Earthquake Engeering and Structural Dynamics*, 29:1–17, 2000.
44. J. R. Field and M. Grigoriu. On the accuracy of the polynomial chaos approximation. *Probabilistic Engineering Mechanics*, 19(1-2):65–80, 2004.
45. G.S. Fishman. *Monte Carlo: concepts, algorithms, and applications*. Springer Series in operations research. Springer-Verlag New York Berlin Heidelberg, 1995.
46. W. B. Fraser and B. Budiansky. The buckling of a column with random initial deflections. *Journal of Applied Mechanics*, 36(2):233–240, 1969.
47. R. Ghanem and R. Kruger. Numerical solution of spectral stochastic finite element systems. *Computer Methods in Applied Mechanics and Engineering*, 129:289–303, 1996.
48. R.G. Ghanem. The nonlinear Gaussian spectrum of log-normal stochastic processes and variables. *Journal of Applied Mechanics*, 66:964–973, 1999.
49. R.G. Ghanem and P.D. Spanos. Spectral stochastic finite-element formulation for reliability analysis. *Journal of Engineering Mechanics*, 117(10):2351–2370, October 1991.
50. R.G. Ghanem and P.D. Spanos. *Stochastic Finite Elements: A Spectral Approach*. Springer Verlag, 1991.
51. G.H. Golub and Van Loan C.F. *Matrix Computations*. The Johns Hopkins University Press, Baltimore, London, 3 edition, 1996.
52. L.L Graham and G. Deodatis. Response and eigenvalue analysis of stochastic finite element systems with multiple correlated material and geometric properties. *Probabilistic Engineering Mechanics*, 16(1):11–29, 2001.
53. M. Grigoriu. Mean-square structural response to stationary ground acceleration. *Journal of Engineering Mechanics*, 107:969–986, 1981.

54. M. Grigoriu. Simulation of nonstationary Gaussian processes by random trigonometric polynomials. *Journal of Engineering Mechanics*, 119(2):328–343, February 1993.
55. M. Grigoriu. *Applied Non-Gaussian Processes: Examples, Theory, Simulation, Linear Random Vibration, and MATLAB Solutions.* Prentice Hall, Englewood Cliffs, 1995.
56. J. Guggenberger and H. Grundmann. Monte carlo simulation of the hysteretic response frame structures using plastification adapted shape functions. *Probabilistic Engineering Mechanics*, 19(1-2):81–91, 2004.
57. P.R. Halmos. *Finite-dimensional vector spaces.* Springer-Verlag New York, Heidelberg, Berlin, second edition, 1974.
58. J. S. Hansen. General random imperfections in the buckling of axially loaded cylindrical shells. *AIAA*, 15(9):1250–12564, September 1977.
59. M. W. Hilburger, V. M. Britt, and M. P. Nemeth. Buckling behavior of compression-loaded quasi-isotropic curved panels with a circular cutout. *International Journal of Solids and Structures*, 38:1495–1522, 2001.
60. M. W. Hilburger and J.H. Starnes. Effects of imperfections on the buckling response of compression-loaded composite shells. *International Journal of Non-Linear Mechanics*, 37(4-5):623–643, 2002.
61. M.W. Hilburger and title = Effects of imperfections of the buckling response of composite shells journal = Thin-Walled Structures volume = 42 number = 3 year = 2004 pages = 369-397 Starnes, J.H.
62. J. M. A. M. Hol. *Shell Design & Analysis System, Usage Notes.* Delft University of Technology, Koiter Institute, The Netherlands, 1996. Version 1.2c.
63. C.S. Hsu. Domain-to-domain evolution by cell mapping. In W. Kliemann and S.N. Namachchivaya, editors, *Nonlinear Dynamics and Stochastic Mechanics*, pages 45–68. CRC Press, Boca Raton, 1995.
64. T.J.R. Hughes. *The Finite Element Method, Linear Static and Dynamic Finite Element Analysis.* Dover Publications, Inc., Mineaola, New York, 2000.
65. J.E. Hurtado. Analyis of one-dimensional stochastic finite elements using neural networks. *Probabilistic Engineering Mechanics*, (44):17–35, 2002.
66. M. Kleiber, editor. *Handbook of computational solid mechanics: Survey and comparison of contemporary methods.* Springer-Verlag, Berlin Heidelberg, 1998.
67. M. Kleiber. Computational stochastic mechanics. *Computer Methods in Applied Mechanics and Engineering*, 168(1-4):1–353, 1999.
68. P.E. Kloeden and E. Platen. *Numerical Solution of Stochastic Differential Equations.* Springer Verlag, Berlin, Heidelberg, New York, 1992.
69. W.T. Koiter. *On the Stability of Elastic Equilibrium.* PhD thesis, Polytechnic Institute Delft, 1945. English Translation, NASA TT F-10, 833 (1967).
70. C.-C. Li and A. Der Kiureghian. An optimal discretization of random fields. Technical Report UCB/SEMM-92/04, Department of Civil Engineering, University of California, Berkeley, California, U.S.A., March 1992.
71. S.C. Lin. Buckling failure analysis of random composite laminates subjected to random loads. *International Journal of Solids and Structures*, 37(51):7563–7576, 2000.
72. Y. K. Lin. *Probabilistic Theory of Structural Dynamics.* Robert E. Krieger Publishing Company, Huntington, New York, 1976.
73. M. Loève. *Probability Theory.* D. Van Nostrand Company, Inc., Princeton, New Jersey, 3 edition, 1963.

74. R. Lorenz. Achsensymmetrische Verzerrungen in dünnwandigen Hohlzylindern. *Zeitschrift des Vereins Deutscher Ingenieure*, 52, 1908.
75. L.D. Lutes and S. Sarkani. *Stochastic Analysis of Structural and Mechanical Vibrations.* Prentice Hall, Upper Saddle River, New Jersey, 1997.
76. MATLAB. *Version 5.3.* The MathWorks Inc., 1999.
77. H.G. Matthies and M. Meyer. Nonlinear Galerkin methods for the model reduction of nonlinear dynamical systems. *Computers and Structures*, 81:1277–1286, 2003.
78. L. Meirovitch. *Computational methods in structural dynamics.* Sijthoff & Noordhoff, Alphen aan den Rijn, The Netherlands; Rockville, Maryland, U.S.A, 1980.
79. B. Miao. Direct integration variance prediction of random response of nonlinear systems. *Computers and Structures*, 46(6):979–983, 1993.
80. P.C. Müller and W.O. Schiehlen. *Linear Vibrations, A theoretical treatment of multi-degree-of-freedom vibrating systems.* Martinus Nijhoff Publishers, Dordrecht, Boston, Lancaster, 1985.
81. G. Muscolino. Linear systems excited by polynomial forms of non-Gaussian filtered processes. *Probabilistic Engineering Mechanics*, 10:35–44, 1995.
82. A. Naess and Johnsen J.M. Response statistics of nonlinear, compliant offshore structures by the path integral solution method. *Probabilistic Engineering Mechanics*, 8(2):91–106, 1993.
83. J. Náprstek. Strongly non-linear stochastic response of a system with random initial imperfections. *Probabilistic Engineering Mechanics*, 14(1-2):141–148, 1999.
84. A.W. Naylor and G.R. Sell. *Linear operator theory in engineering and science.* Number 40 in Applied mathematical sciences. Springer-Verlag New York Inc., 2 edition, 2000.
85. D. E. Newland. *An Introduction to Random Vibration, Spectral & Wavelet Analysis.* Longman Scientific & Technical, Longman Group Limited, Longman House, Burnt Mill, Harlow Essex CM20 2JE, England, 3 edition, 1965.
86. Y. Ohtori and B.F. Spencer. Semi-implicit integration algorithm for solution of nonlinear stochastic vibration problems. In *8th ASCE Speciality Conference on Probabilistic Mechanics and Structural Reliability*, volume CD-ROM:PMC2000-323, 2000.
87. G. V. Palassopoulos. Buckling analysis and design of imperfection-sensitive structures. In A. Haldar, A. Guran, and B. M. Ayyub, editors, *Uncertaintiy Modeling in Finite Element, Fatigue and Stability of Systems*, volume 9, pages 311–356. World Scientific Publishing Company, 1997. Series on Stability, Vibration and Control of Systems Series B.
88. K.C. Park and G.M. Stanley. A curved C^0 shell element based on assumed natural-coordinate strains. *Journal of Applied Mechanics*, 53:278–290, 1986.
89. B.-F. Peng and J.P. Conte. Closed-form solutions for the response of linear systems to fully nonstationar earthquake excitations. *Journal of Engineering Mechanics*, 124:684–694, 1998.
90. PERL. *Version 5.005-02.* L. Hall, 1998.
91. F. Pfeiffer. *Einführung in die Dynamik.* Teubner Studienbücher Mechanik, Stuttgart, 2nd edition, 1992. Leitfäden der angewandten Mathematik und Mechanik; Bd. 65.

92. K. K. Phoon, H. W. Huang, and S. T. Quek. Comparison between karhunen-loève and wavelet expansions for simulation of gaussian processes. *Computers and Structures*, 82(13-14):985–991, 2004.
93. K. K. Phoon, S. P. Huang, and S. T. Quek. Simulation of second-order processes using karhunen-loève expansion. *Computers and Structures*, 80(12):1049–1060, 2002.
94. K.K Phoon, S.P. Huang, and S.T. Quek. Implementation of Karhunen-Loeve expansion for simulation using a wavelet-Galerkin scheme. *Probabilistic Engineering Mechanics*, (17):293–303, 2002.
95. F. Poiron. Numerical simulation of homogeneous non-gaussian random vector fields. *Journal of Sound and Vibration*, 160(1):25–42, 1993.
96. D.C. Polidori, J.L. Beck, and C. Papadimitriou. A new stationary PDF approximation for nonlinear oscillators. *International Journal of Non-Linear Mechanics*, 35:657–673, 2000.
97. H. J. Pradlwarter and G. I Schuëller. Stochastic response evaluation of large FE-models with non-linear hysteretic elements by complex mode synthesis. In P. D. Spanos, editor, *Proceedings Third International Conference on Computational Stochastic Mechanics*, pages 551–558, Thera-Santorini, Greece, 1999. Balkema, Rotterdam.
98. H.J. Pradlwarter. Deterministic integration algorithms for stochastic response computations of FE-systems. *Computers and Structures*, 80(18-19):1489–1502, 2002.
99. H.J. Pradlwarter and G.I. Schuëller. Equivalent linearization - a suitable tool to analyse MDOF-systems. *Probabilistic Engineering Mechanics*, 8:115–126, 1993.
100. H.J. Pradlwarter, G.I. Schuëller, and C.A. Schenk. A computational procedure to estimate the stochastic response of large non-linear FE-models. *Computer Methods in Applied Mechanics and Engineering*, 192(7-8):777–801, 2003.
101. M.B Priestly. Power spectral analysis of nonstationary random processes. *Journal of Sound and Vibration*, 6(1):86–97, 1967.
102. C. Proppe, H.J. Pradlwarter, and G.I. Schuëller. Equivalent linearization and Monte Carlo simulation in stochastic dynamics. *Probabilistic Engineering Mechanics*, 18(1):1–15, 2003.
103. C. C. Rankin. Personal communication. Institute of Engineering Mechanics, Leopold Franzens University, Austria, EU, 2000.
104. C. C. Rankin, F. A. Brogan, W. A. Loden, and H. D. Cabiness. *STAGS (STructural Analysis of General Shells) User Manual, LMSC P032594, Version 3.0.* Lockheed Martin Missiles and Space Co., Inc., Palo Alto, CA, USA, 1998.
105. S.O. Rice. Selected papers on noise and stochastic processes. volume 23, chapter Mathematical Analysis of Random Noise, pages 133–294. Dover, 1954. paper originally appeared in two parts in the Bell System Technical Journal in Vol. 23, July 1944 and in Vol. 24, January 1945.
106. E. Riks. Some computational aspects of the stability analysis of nonlinear structures. *Computer Methods in Applied Mechanics and Engineering*, 47:219–259, 1984.
107. E. Riks. Progress in collapse analyses. *Journal of Pressure Vessel Technology*, 109:33–41, 1987.
108. E. Riks, C. C. Rankin, and F. A. Brogan. The numerical simulation of the collapse process of axially compressed cylindrical shells with measured imperfec-

tions. Technical Report LR-705, TU Delft, Faculty of Aerospace Engineering, December 1992.
109. J.B. Roberts and P.D. Spanos. *Random Vibration and Statistical Linearization.* Wiley & Sons, Chichester, New York, 1990.
110. J. Roorda. Some statistical aspects of the buckling of imperfection-sensitive structures. *Journal of Mechanics and Physics of Solids*, 17(1):111–123, 1969.
111. S. Sakamoto and R. Ghanem. Simulation of multi-dimensional non-gaussian non-stationary random fields. *Probabilistic Engineering Mechanics*, 17(2):167–176, 2002.
112. C. A. Schenk. Some aspects of simulation of stochastic processes and fields. Institute of Engineering Mechanics, Leopold-Franzens University, Innsbruck, Austria, EU, 2001. Int. Work. Rep. 42-01.
113. C. A. Schenk and G. I. Schuëller. Buckling analysis of cylindrical shells with random boundary and geometric imperfections. In R. Corotis, G. I. Schuëller, and M. Shinozuka, editors, *Proc. of 8th International Conference on Structural Safety and Reliability, ICOSSAR' 01*, pages 137–143. Swets, Rotterdam, 2002.
114. C. A. Schenk and G. I. Schuëller. Buckling analysis of cylindrical shells with random geometric imperfections. *International Journal of Non-Linear Mechanics*, 38(7):1119–1132, 2003.
115. C. A. Schenk, G. I. Schuëller, and J. Arbocz. On the analysis of cylindrical shells with random imperfections. In M. Papadrakakis, A. Samartin, and E. Oñate, editors, *Proceedings of the Fourth International Colloquium on Computation of Shell & Spatial Structures*, Chania - Crete, Greece, June 2000. IASS-IACM 2000. CDRom.
116. C.A. Schenk and G.I. Schuëller. Buckling analysis of cylindrical shells with cutouts including random geometric imperfections. In H.A. Mang, F.G. Rammerstorfer, and J. Eberhardsteiner, editors, *Proceedings of the Fifth World Congress on Computational Mechanics (WCCM V).* IACM, Vienna University of Technology, Austria, 2002. http://wccm.tuwien.ac.at, ISBN 3-9501554-0-6.
117. G. I. Schuëller and C. E. Brenner. Advances in Building Materials Science. In A. Gerdes, editor, *Stochastic Finite Elements - A Challenge for Material Science*, pages 259–281. Aedificatio Publ., 1996.
118. G. I Schuëller and H. J. Pradlwarter. On the stochastic response of nonlinear FE models. *Archive of Applied Mechanics*, 69:765–781, 1999.
119. G.I. Schuëller. Structural reliability - recent advances - Freudenthal Lecture. In *Proceedings of the 7th International Conference on Structural Safety and Reliability (ICOSSAR '97)*, pages 3–35. A.A. Balkema Publications, Rotterdam, The Netherlands, 1998.
120. G.I. Schuëller. Computational stochastic mechanics - recent advances. *Computers and Structures*, 79(22-25):2225–2234, 2001.
121. G.I. Schuëller, M.D. Pandey, and H.J. Pradlwarter. Equivalent linearization (EQL) in engineering practice for aseismic design. *Journal of Probabilistic Engineering Mechanics*, 9:95–102, 1994.
122. G.I. Schuëller, H.J. Pradlwarter, and C.A. Schenk. Nonstationary response of large linear FE-models under stochastic loading. *Computers and Structures*, 81(8-11):937–947, 2003.
123. G.I. Schuëller, H.J. Pradlwarter, M. Vasta, and N. Harnpornchai. Benchmark-study on non-linear stochastic structural dynamics. In N. Shiraishi, M. Shinozuka, and Y.K. Wen, editors, *ICOSSAR'97 (7th Intern. Conf. on Structural*

Safety and Reliability, Kyoto, Nov. 1997, pages 355–361. A.A. Balkema Publ., Rotterdam, The Netherlands, 1998.

124. G.I. Schuëller (Ed). A state-of-the-art report on computational stochastic mechanics. *Probabilistic Engineering Mechanics*, 12(4):197–321, 1997.
125. G.I. Schuëller (Ed). Computational stochastic mechanics and reliability analysis (special issue). *Computer Methods in Applied Mechanics and Engineering*, 194(12-16):1251–1795, 2005.
126. G.I. Schuëller (Ed). Uncertainties in structural mechanics and analysis - computational methods (special issue). *Computers and Structures*, 2005. to appear.
127. E. Senocak and A.M. Waas. Optimally reinforced cutouts in laminated circular cylindrical shells. *International Journal of Mechanical Sciences*, 38(2):121–140, 1996.
128. M. Shinozuka. Monte Carlo solution of structural dynamics. *Computers and Structures*, 2(5+6):855–874, 1972.
129. M. Shinozuka. Digital simulation of random processes in engineering mechanics with the aid of FFT technique. In S.T. Ariaratnam and H.H.E. Leipholz, editors, *Stochastic Problems in Mechanics*, pages 277–286, Waterloo, 1974. University of Waterloo Press.
130. M. Shinozuka. Stochastic fields and their digital simulation. In G. I. Schuëller and M. Shinozuka, editors, *Stochastic Methods in Structural Dynamics*, pages 93–133. Martinus Nijhoff Publishers, 1987.
131. M. Shinozuka and G. Deodatis. Response variability of stochastic finite element systems. *Journal of Engineering Mechanics*, 114(3):499–519, 1988.
132. M. Shinozuka and G. Deodatis. Simulation of stochastic processes by spectral representation. *Applied Mechanics Reviews*, 44(4):191–204, 1991.
133. M. Shinozuka and G Deodatis. Simulation of multi-dimensional Gaussian stochastic fields by spectral representation. *Applied Mechanics Reviews*, 49(1):29–53, 1996.
134. M. Shinozuka and C Jan. Digital simulation of random processes and its applications. *Journal of Sound and Vibration*, 25(1):111–128, 1972.
135. M. Shinozuka and F. Yamazaki. Stochastic finite element analysis: an introduction. In S. T. Ariaratnam, G. I. Schuëller, and I. Elishakoff, editors, *Stochastic Structural Dynamics*, pages 241–291. Elsevier Applied Science, 1988.
136. H. Shuping. *Simulation of random processes using Karhunen-Loéve expansion.* PhD thesis, The National University of Singapore, 2001.
137. B.N. Singh, D. Yadav, and N.G.R. Iyengar. Initial buckling of composite cylindrical panels with random material properties. *Composite Structures*, 53(1):55–64, 2001.
138. B.N. Singh, D. Yadav, and N.G.R. Iyengar. Stability analysis of laminated cylindrical panels with uncertain material properties. *Composite Structures*, 54(1):17–26, 2001.
139. A.W. Smyth and S.F. Masri. Nonstationary response of nonlinear systems using equivalent linarization with a compact analytical form of the excitation process. *Probabilistic Engineering Mechanics*, 17:97–108, 2002.
140. C. Soize. Stochastic linearization method with random parameters for SDOF nonlinear dynamical systems: prediction and identification procedures. *Probabilistic Engineering Mechanics*, 10(3):143–152, 1995.
141. T.T. Soong and M. Grigoriu. *Random Vibration of Mechanical and Structural Systems.* Prentice Hall, Englewood Cliffs, New Jersey, 1993.

142. P. D. Spanos and B. A. Zeldin. Monte Carlo treatment of random fields. *Applied Mechanics Reviews*, 51(3):219–237, 1998.
143. B.F. Jr. Spencer and L.A. Bergman. On the numerical solution of the Fokker-Planck equation for nonlinear stochastic systems. *Nonlinear Dynamics*, 4:357–372, 1993.
144. J.H. Starnes. The effects of cutouts on the buckling of thin shells. In Y. C. Fung and E. E. Sechler, editors, *Thin-Shell Structures: Theory, Experiment and Design*, pages 289–304. Prentice-Hall, 1974.
145. J.H. Starnes and M.W. Hilburger. The effects of reinforced cutouts on the buckling of composite shells. In H.A. Mang, F.G. Rammerstorfer, and J. Eberhardsteiner, editors, *WCCM V, Fifth World Congress on Computational Mechanics*. IACM, 2002.
146. G. Stefanou and M. Papadrakakis. Stochastic finite element analysis of shells with combined random material and geometric properties. *Computer Methods in Applied Mechanics and Engineering*, 193(1-2):139–160, 2004.
147. A. Steindl and H. Troger. Methods for dimension reduction and their application in nonlinear dynamics. *International Journal of Solids and Structures*, 38(10-13):2131–2147, 2001.
148. Y. Suzuki and R. Minai. Application of SDEs to seismic reliability analysis of hysteretic structures. *Probabilistic Engineering Mechanics*, 3(1):43–52, 1988.
149. A. Tafreshi. Buckling and post-buckling analysis of composite cylindrical shells with cutouts subjected to internal pressure and axial compression loads. *International Journal of Pressure Vessels and Piping*, 79(5):351–359, 2002.
150. S. P. Timoshenko. *Theory of Elastic Stability.* McGraw-Hill Book Co., New York, 1936.
151. C.W.S. To. Direct integration operators and their stability for random response of multi-degree-of-freedom systems. *Computers and Structures*, 30(4):865–874, 1988.
152. Y. Tsompanakis, V. Papadopoulos, N.D. Lagaros, and M. Papadrakakis. Reliability analysis of structures under seismic loading. In H.A. Mang, F.G. Rammerstorfer, and J. Eberhardsteiner, editors, *Proceedings of the Fifth World Congress on Computational Mechanics (WCCM V).* IACM, Vienna University of Technology, Austria, 2002. http://wccm.tuwien.ac.at, ISBN 3-9501554-0-6.
153. E. H. Vanmarcke. *Random Fields: Analysis and Synthesis.* MIT Press, Cambridge, Massachusetts, London, England, 1983.
154. M. Vannucci, A. Moro, and P.D. Spanos. Wavelets in random processes representation. In D.M. Frangopol and M.D. Grigoriu, editors, *Probabilistic Mechanics and Structural Reliability*, pages 672–675, 1996.
155. W.V. Wedig. Iterative schemes for stability problems with non-singular fokker-planck equations. *International Journal of Non-Linear Mechanics*, 31(5):707–715, 1996.
156. C. Wen, R. Wen-Min, and Z. Wei. Buckling analysis of ring-stiffened cylindrical shells with cutouts by mixed method of finite strip and finite element. *Computers and Structures*, 53(4):811–816, 1994.
157. Y.K Wen. Equivalent linearization for hysteretic systems under random excitation. *Journal of Applied Mechanics*, 47:1–20, 1980.
158. S.F. Wojtkiewicz and L.A. Bergman. Numerical solution of a four-dimensional fokker-planck equation. In K.J. Bathe, editor, *Proceedings of the First MIT Conference on Computational Fluid and Solid Mechanics*, volume 1, Amsterdam, Boston, June 2001. Elsevier. CDRom.

159. D. Yadav and N. Verma. Buckling of composite circular cylindrical shells with random material properties. *Composite Structures*, 37(3-4):385–391, 1997.
160. F. Yamazaki and M. Shinozuka. Digital generation of non-Gaussian stochastic fields. *Journal of Engineering Mechanics*, 114(7):1183–1197, 1988.
161. J.N. Yang. Simulation of random envelope processes. *Journal of Sound and Vibration*, 25(1):73–85, 1972.
162. J.N. Yang. On the normality and accuracy of simulated random processes. *Journal of Sound and Vibration*, 26(3):417–428, 1973.
163. C.-H. Yeh and Y.K. Wen. Modeling of nonstationary ground motion and analysis of inelastic structural response. *Structural Safety*, 8:281–298, 1990.
164. M.-K. Yeh, M.-C. Lin, and W.-T. Wu. Bending buckling of an elastoplastic cylindrical shell with a cutcut. *Engineering Structures*, pages 996–1005, 1999.
165. B.A. Zeldin and P.D. Spanos. Random field simulation using wavelet bases. *Journal of Applied Mechanics*, 63(4):946–952, 1996.

Index

Printing: Krips bv, Meppel
Binding: Stürtz, Würzburg